Clockmaking
In The Vale of Clwyd

Paul Parker

*For
Carole & her Uncle Vic,
without whom this book
might not have been written*

Published by the Author

© Paul Parker, 1993
First published 1993

All rights reserved. No part of this publication may be reproduced, in any form or by any means, without permission from the Publisher.

Printed in Great Britain
by Studio 365, Mold, Clwyd.

ISBN 0 9522617 0 7

INTRODUCTION

For reasons which will become obvious later, it was easy to decide that the period which this book would cover would start in 1605 and, although the choice of 1900 as cut-off point was slightly arbitrary and has not been adhered to with total exactitude, it is a good round number and few, if any, will object to the occasional incursion into the 20th century. On the other hand, it was not so easy to define the area to be covered because, although the Vale of Clwyd is bordered on one side by the hills of the Clwydian Range and on the other by the Denbigh Moors, and although the River Clwyd flows along its entire length, there is no unanimous agreement on its precise geographical limits. At its northern end it blends imperceptibly into the coastal strip which runs from Abergele to Prestatyn and well beyond, and opinions are divided as to whether or not Corwen should be included in the South. Moreover, until the local government reorganisation in 1974, the parishes of Rhuddlan, Rhyl and St. Asaph were in Flintshire, whereas the rest of the Vale was in Denbighshire. There is no record of clockmaking or clockmakers in some parts of the Vale and so this book concentrates on the parishes of Denbigh, Henllan, Rhuddlan, Rhyl, Ruthin and St. Asaph. The River Clwyd runs through all of these parishes except Henllan. However, the parish of Henllan used to extend into the borough of Denbigh, so that houses on the southern side of Henllan Street were in Denbigh parish, whereas the houses on the opposite side were in Henllan parish. Research thus led me inescapably into Henllan. I have also referred occasionally to Bodfari, whose church had a clock in the 18th century and whose parish overlooks the Vale of Clwyd and Denbigh.

Of the Vale towns, Rhyl is a relatively modern invention. Until the 19th century it was no more than a scattering of fishermen's cottages but with the building of the Chester to Holyhead railway it grew into a busy and prosperous sea-side resort. Rhuddlan, on the other hand, was once an important borough as its "Parliament House" and imposing castle testify. By the 18th century, however, it had become a sleepy country town. St. Asaph was also a quiet country town, although its status as cathedral city may have given it a certain moral authority denied to the larger towns of Denbigh and Ruthin, between whom there has always been (and continues to be) a healthy rivalry. During the period which we shall be considering, Denbigh generally held the upper hand and had done so for a long time. The Earl of Leicester's plan for it to replace St. Asaph as the cathedral city reflected, amongst other things, its leading role as a thriving and influential community, which the Elizabethan historian Humphrey Llwyd[1], who lived in Denbigh, described in these words:

> "... being the chief town of the shire, standing in the very middle of the country, it is a great market town, famous and much frequented with wares and people from all parts of North Wales. The indwellers have the use of both tongues, and, being endued by the kings of England with many privileges and liberties, are ruled by their own laws."

As far as clockmaking is concerned, Denbigh appears to have been predominant until the second half of the 19th century as the four charts (figure 1) clearly show. The dark shade indicates clockmakers for whom we have more than one reference whereas the lighter shade shows the number of makers who only seem to be recorded once. However, a note of caution must be sounded.

---

[1] Sometimes known by the anglicised version of his name, Lloyd.

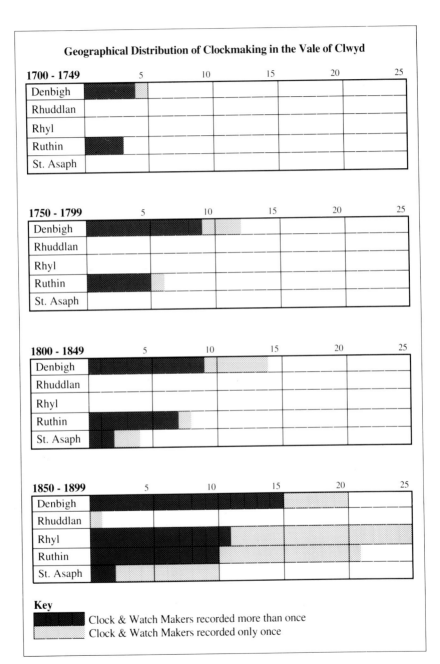

[Figure 1]

Firstly, a clockmaker may only have been recorded once and yet have made a long and full contribution to clockmaking in his community. The majority of our evidence comes after all from parish registers. If a man did not marry locally (or did not marry at all) and had no children (or did not have them baptised) then we may have only his burial entry to show that he existed, despite the fact that he could have been a working clockmaker for all of his working life.

Secondly, since there was no standardised form for recording baptisms, marriages and burials, the information supplied by rectors in different parishes, and different rectors in the same parish, varies widely. In Henllan, St. Asaph and Rhuddlan, the rectors usually identified their parishioners by their place of residence. A David Davies of Llan was thus distinguishable from a David Davies of Bannister Isaf. However, Denbigh and Ruthin rectors almost invariably noted their parishioners' job, profession or status, so that a John Jones, Labourer would not be confused with a John Jones, Perriwig Maker. This latter method is obviously far more helpful for our research. After all, unlikely as it is, both David Davies from Llan and David Davies from Bannister Isaf might just conceivably have been clockmakers and it is certainly possible that one of them might have been a farmer who sometimes turned his hand to clockmaking during the winter months. This was common in the Black Forest in Germany and, as we shall see later, we have evidence that it happened in the Vale of Clwyd too.

The charts are therefore not infallible. Nevertheless, the picture which they show of clockmaking in the four half-centuries covered is consistent with the populations of the towns they cover, and is thus probably a fair reflection of horological activity.

With regard to the text, everything between quotation marks has been checked and re-checked with the result that I can vouch for the absolute accuracy of all the quotations, down to the last capital letter; as a consequence, I have not followed the convention of placing [sic] after an unusual spelling except when that unorthodox spelling or word is used in normal text outside quotation marks. Also, there are a number of references to pre-decimal money. Rather than place the modern equivalent inside brackets after every reference to shillings and pence, I hope that it will be sufficient to remind readers that six pre-decimal pence were the equivalent of two and a half new pence, that one shilling was the equivalent of five new pence and that one guinea was the equivalent of one pound and five new pence.

During my research I have had reason to be grateful to a large number of people for their help and advice. There are far too many to mention all of them individually, but in particular I must thank: Mr. R.M. Owen of Denbigh for his excellent advice at all stages·of this project; Mr. Gareth Jones, Clock and Watch Conservator at the Welsh Folk Museum, for his encouragement and technical suggestions; Mr. David Griffith of Bradford-upon-Avon for his invaluable information about his ancestors who made clocks in Denbigh for over a hundred years; Mrs. Pauline Roberts and her son Mr. Christopher Keepfer Roberts for their extremely helpful background information about the Keepfer family; all the staff at the Clwyd Record Offices in Ruthin and Hawarden for being unfailingly helpful and exceedingly patient; Mr. Ronald Thompson and Mr. Duncan Waldron for allowing me to use their photographs; all the families who have most kindly allowed me to invade their homes, albeit temporarily, to take pictures of their treasured clocks; my colleague Miss C.P.Whitaker for her eagle-eyed proofreading; and finally Carole, who has seen to it that I persevered when things were not going according to plan, who has fortified me with countless cups of coffee, whose advice has always been sound, and who has not complained at the many hours I have sat hunched over my mounds of documents and the word processor.

As we saw in the introduction, Denbigh was a town of considerable importance in Tudor and Stuart times and certainly the most prosperous of the Vale towns. A number of rich and influential families such as the Myddletons and the Cloughs were well established there and maintained close links with London. There can be little doubt that at least some of these families owned domestic clocks such as the lantern clocks that were reasonably widespread among wealthy households during this period. However, there appears to be no record of their purchase or ownership, and they were probably so scarce in the Vale that there was no call for the services of a local clockmaker to service or repair them.

Nor is there any evidence of public clocks in any of the most likely sites before the beginning of the 17th century. It is generally agreed that mechanical clocks were first constructed for use in monasteries to help regulate the daily worship and the Vale of Clwyd has a number of sites which might well have had their clocks from the 15th or 16th centuries. However, we have to wait until 1605 for our first reference to a clock or clockmaker in the Vale. Not surprisingly, Denbigh is the site of the clock. The reference is in the Corporate Records and below I provide a transcript of the original (figure 2):

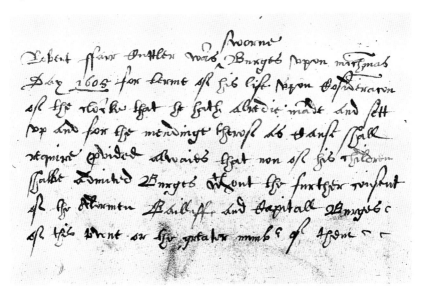

[Figure 2 - By kind permission of the Clwyd Record Office]

"Robert ffarr Cuttler was sworne Burgess upon Mich'mas Day 1605 for term of his life upon considera on of the clocke that he hath alredie made and sett up and for the mendinge therof as cause shall require provided alwaies that non of his children shalbe admitted Burgess without the further consent of the Aldermen Baillifs and Capitall Burgesses of this towne or the greater number of them."

John Williams in his book "Ancient and Modern Denbigh", which was published in 1856 and reprinted in 1989, has a substantially different version of this entry[2], not only because he has added the appropriate punctuation, but also because he has included a passage not to be found in the original:

> "Robert ffarr, Cutler, was sworne burgess vpon Mich'mas Day, 1605, for term of his life, vpon consider on of the clocke that he hath alredie made and sett up, and for the mendinge therof as cause shall require, &c., *which clocke hee valued to be worth £V, and which he was bound to set, tend, and repair duringe his naturall life.*" [my italics]

John Williams' version of the entry appears to combine the shortened original plus information from another source and, since he did not leave a precise reference for that source, it has not proved possible to trace this part of the contract.

With or without the John Williams' additional information, this is a tantalising entry, both for what it says and for what it does not say. As so often, it is a great pity that the clerk did not realise that a full technical description of the clock and its construction would turn out to be helpful to researchers in the 20th century, as well as some notes on how Robert Farr went about his task. However, we must not be unduly ungrateful because, even if he was not able to foresee the needs of generations nearly four centuries later, he did provide us with a considerable amount of useful information.

To begin with, he clearly states that Robert Farr "made and sett up" the clock, and that he was a cutler. This leaves us in no doubt that Robert Farr built the clock himself, and that he was as well equipped as any other man in the community to do so, since by trade he was a metalworker. Presumably he would have belonged to the Company of Hammermen, a guild established in Denbigh since medieval times whose membership included metalworkers of all descriptions and possibly some allied trades. Many clocks in the United Kingdom were built by local craftsmen, most often by blacksmiths, since the main qualifications to do the job were on the one hand the ability to work with iron and on the other hand the spirit of adventure - or perhaps the lure of financial advantage. The overwhelming majority of them probably built only the one clock in their whole lives, although some of them almost certainly went on to make clockmaking into an important, and in some cases the only, part of their trade.

There was, however, not only adventure or monetary advantage in the building of a clock. The clock then had to be looked after and repaired while the clockmaker remained alive, and this was no mean undertaking. To appreciate this, we need to understand the horological revolution of the 17th century.

The year 1657 is a watershed in horology, for this year saw the successful harnessing of the pendulum to the clock. Before 1657, the accuracy of clocks was a hit and miss affair which depended on a variety of factors. In particular, clocks speeded up in winter and slowed down in summer, and their speed fluctuated as the wheels turned, because some of the gear teeth would mesh more smoothly than others; this was especially noticeable in the more crudely built clocks, since making a workable train

---

[2] reprinted 1989 by Adlib Books, Combe Park, Bath. p. 282

of gears would have been the most difficult part of the operation for a first-time clockmaker and it was not unusual for a clock to be wildly inaccurate only a few hours after being set. The only saving grace was that clocks had only one hand, the hour hand, and that while they were within about a quarter of an hour of the correct time no great harm was done.

The introduction of pendulum-controlled clocks revolutionised their accuracy, all the more so since, within the space of a further 20 years, a new escapement was developed (this time by British physicists as opposed to the Dutchmen who had harnessed the pendulum). The anchor escapement, in conjunction with the pendulum, meant that it was commonplace for clocks to keep time to within a minute a week, and frequently to within a few seconds. One sign that a clock was of the new, accurate type was the presence of a minute hand and often a seconds hand as well and this led to a nationwide re-education in telling the time. Clocks made in the late 17th and early 18th centuries were designed to be read by people familiar with either the one-handed or two-handed variety.

To return to Robert Farr and his clock, the alert reader will already have realised the implications of our brief lesson. His clock, built in or shortly before 1605, did not have the benefits of the pendulum, let alone the anchor escapement. It was controlled by a foliot, a metal arm which swung horizontally first in one direction, then the other, in a fairly haphazard rhythm. In summer, to counteract the slowing down caused by metal expansion, small weights hung on the foliot could be moved inwards towards the central pivot; in winter, the opposite would need to be done; and, since the British weather is notoriously changeable, cold spells in the summer or warm spells in winter would cause daily timekeeping problems. Alternatively, the main driving weight could be varied by adding extra small weights or removing them as required.

However, this would have been only the beginning of Robert Farr's problems. How was he to know whether the clock was slow or fast? He had no Greenwich signals, no talking clock, no radio time checks. His only means of comparison was with the natural time as indicated by the sun and he would have depended on a sundial, perhaps hand-held, to set the clock to time. Indeed, he could have had a sundial-literate colleague to shout out the time to him from below as he worked up inside the county hall. Moreover, the sun could not be relied upon to shine every day and, by the end of a couple of days of cloudy weather, Robert Farr might well be less than popular with anyone who had any need to keep time to within the odd hour or two. In fact, it is doubtful whether this applied to many people; after all, the main reason for wanting a public clock was so that everybody was tuned in to the same time, whether it was accurate or not, and the system, for all its faults, was probably felt to be an improvement on having no publicly agreed time at all.

This is of course borne out by the decision of the aldermen and capital burgesses to admit Robert Farr to the freedom of the town once he had completed the clock. Being a burgess conferred status and financial advantages on the holder and to become a burgess one had broadly speaking either to be influential or likely to be of use to the town. The freedom of the town was often offered free of charge to men of wealth; others usually had to pay for the privilege. According to John Williams, "foriners" -

that is to say by and large Welshmen - were charged £5 for their admission[3] and it is probably no coincidence that, again according to John Williams, Robert Farr's clock was valued at "£V". Although freedom of the town could not be passed on to children who had already been born, it was passed on to further children, who therefore automatically inherited the right to trade in Denbigh. This privilege, in a prosperous community, was clearly of considerable value.

It should be pointed out that he would be setting the clock to Denbigh local time, not Greenwich mean time, and that his clock, when accurate by the sundial, would not only be nearly 14 minutes later than London time[4], but would also be up to a quarter of an hour fast or slow in comparison with today's accepted system because of the difference between what we shall call "sundial time" and "clock time". Fortunately, none of this can have mattered at all to anybody in Denbigh in 1605, just as adding "leap" seconds nowadays is of only academic interest to someone waiting for the bus to Ruthin.

Having built his clock and developed a routine for winding it and setting it (which probably involved more than one visit each day), Robert Farr still needed to maintain it in working order and a cursory inspection of the churchwardens' accounts from the 17th or 18th centuries relating to any church which owned a clock will immediately reveal the quantity of "oyle" as well as "wyre" or "wyer" that was required to keep a clock going. Oil made from animal fats[5] degraded very quickly, losing its lubricating qualities and turning into a thick paste which clogged up the working parts and brought the clock to a grinding halt. This paste would also have absorbed large amounts of dust, dirt and metal particles which would wear down the metal working surfaces, thereby necessitating further repairs. Much damage could be caused by adding oil to the paste; this allowed the clock to work temporarily, but the dust and metal particles would be doing the mechanism no good at all, merely shortening the working life of the components. Ideally, Robert Farr's clock would have had to be completely dismantled, cleaned and reassembled before being oiled; cutting corners would have proved more costly and time-consuming in the long run. However, it is more than likely that these corners were very frequently cut.

It may have been noticed that the Corporation records do not specify where the clock was sited and there were certainly two public clocks in Denbigh by the 18th century. One was in St. Hilary's chapel and the other in the Town Hall[6]. The fact that it is the Borough of Denbigh which is recording the event makes it virtually certain that the clock we are discussing is the Town Hall clock.

Having seen the probable circumstances surrounding Robert Farr's foray into clockmaking, we can now only guess at how he designed his clock. In 1605 there was no government training centre in Denbigh, nor was there a correspondence course organised by the British Horological Institute. In any case, we know that Robert Farr

---

[3] Op. cit., pp. 106-107
[4] Denbigh lies 3° 25' west of the Greenwich meridian.
[5] Bodfari churchwardens' accounts refer to "Sweet Oyle to ye Clock comp^d of Hog's Greass".
[6] This is known nowadays as the County Hall and is the building which houses the Museum and Library. I have referred to this building throughout as the Town Hall because the documents use this title.

was not a qualified clockmaker but a cutler. Three theories, therefore, can be suggested. The first, that he was a self-taught genius who designed and built a clock with no outside help of any sort. The second, that the design and construction of turret clocks was explained to him by someone of experience. The third, that he copied an already existing clock.

It does not take much thought to see that the third theory is the most likely, although it is obvious that, when copying another clock, he would have been well advised to discuss the possible pitfalls with whoever built or looked after the clock he was copying. As we have already seen, the most difficult part of the operation would have been to construct gear wheels and pinions which meshed as accurately as possible but, by copying another clock that was already working satisfactorily, he would only have needed to use exactly the same size wheels with the same number of teeth. Where he found a clock to copy we do not know but, if there was no clock in the Vale for him to copy, he would have needed to go no further than Chester or Wrexham to find a suitable model.

If this theory is indeed correct, and it seems highly probable, then Denbigh's first public clock was not at the cutting edge of horological research; it was, on the contrary, a copy of an already old design, built by a local craftsman with almost certainly no previous experience of clockmaking and relying on the advice of another local metalworker, probably also with little or no horological training.

In the event, Robert Farr's clock did very well. We can only speculate as to how good a timekeeper it was and how reliable it proved to be. Certainly, it is probable that it caused problems, since we read in the Corporation Records in 1671 that Robert Maurice, a baker, paid £2.8s.6d towards "ye repayring ye clocke next fayre day." However, its maker produced a clock that - with constant attention and liberal quantities of oil and replacement spares - lasted until the early 19th century when it was finally replaced.

II

If the story of 17th century clockmaking in the Vale of Clwyd is confined to one Denbigh clock and some speculation, the 18th century saw the establishment of the first of three clockmaking families in Denbigh, with the arrival of the Minshulls, and the growth of clockmaking in Ruthin.

Two Denbigh parish registers overlap briefly at the beginning of the century and in the first we read:

> "Katherina filia Henrici Minshull Clockmaker et ejus uxoris Maria nata fuit 1mo Die Julii et baptizata 15° Die ejusdem mentis 1705."

followed by the same information in the second:

> "Katherine ye Daughter of Henry Minshall[7] and Mary his wife was born ye jst of July and was baptized ye 15 instant 1705."

We do not know where Henry came from; it is possible that he was related to the clockmaking Minshulls of Nantwich in Cheshire, or that he was related to the Minshulls of Stoke and Chester who were admitted to the freedom of Denbigh in 1681[8]. Be that as it may, we do know a few things about him. To begin with, he is described as a clockmaker in the first of these two entries. This means that he had completed a 7-year apprenticeship probably from the age of 12, and it was extremely common for young men to marry the moment they were properly qualified since as apprentices they were not allowed to do so. We also know that it is extremely likely that he and his wife already had one son when they arrived in Denbigh, since a John Minshull followed in Henry's footsteps, the first reference to him being in 1722 when he was already a clockmaker . Again applying the rule of the 7-year apprenticeship from 12 to 19, it would be reasonable to assume that John was born some time between 1700 and 1703, which is consistent with his having died in 1778.

Henry was to have another daughter, Rachell, in April 1709 - on this parish register entry Henry is called Harry - but no further children are recorded after this; nor do we know what happened to Katherine and Rachell except that, in the absence of any burial entries in the parish register, they appear to have survived childhood.

The establishment of a full-time and fully qualified clockmaker in a town of Denbigh's importance is of course not in the least surprising bearing in mind that, as we have already established, the second half of the 17th century was an era of quick and spectacular improvements in the accuracy of clocks. As clocks became more accurate and so more useful, they automatically became more desirable and clock ownership suddenly came to be considered a necessity for many more households. In addition, ownership of a clock carried with it the status associated with the most up-to-date scientific advances of the day. Until the advent of the motor car a clock was the most technologically sophisticated product that the average family could possess; not to possess one was not to share in the exciting scientific progress which would shortly lead to mastery of the seas through mastery of time and therefore of navigation. Even

---

[7] Variations on Minshull include Minshul, Mynshull, Mynshul, Minshuls, Minshall, Minshal and Minchor.
[8] John Williams: Records of Denbigh and its Lordships, p. 140

the most humble of 30-hour single-handed "cottage" clocks was part of the scientific revolution, in the same way as owning a family saloon has a link with the glamour and technological advances of Formula 1 car racing.

Only one of Henry Minshull's clocks has so far come to light, although there must be others still keeping time two and a half centuries after they were made. The one example I have seen is an 8-day longcase clock which is not in its original case, and whose movement has a number of modern replacement parts. However, the basic movement is original, as is the dial.

The style of the dial is consistent with the early years of the 18th century, perhaps 1710 or 1715. As I mentioned in chapter 1, early 18th century clocks were intended to be read not only by people who were familiar with the newer two-handed dials but also by those who were either happier with the older one-handed dials or who had not learned to read the newer type.

It is instantly obvious that, in the case of a clock which has only one hand - that is to say the hour hand - the hours can be divided quite reasonably into four quarters by means of the four divisions engraved on the *inner* edge of the chapter ring and the top half of the diagram in figure 3 shows the chapter ring markings for such a clock. When the hand points directly at one of the lines, it is exactly the hour, the half hour or one of the quarters. Equally obviously, it is not too difficult to see with some degree of accuracy when the single hand is halfway between two of the lines; in our example, the time is just between half past twelve and a quarter to one, in other words twenty-two and a half minutes to one. In this way a single handed clock can be used to determine the time to the nearest seven and a half minutes.

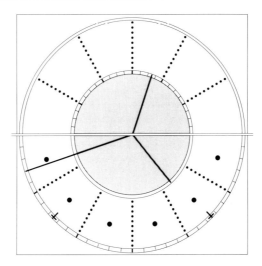

Figure 3: Chapter ring markings for (top) one-handed clock, (bottom) two-handed clock. The time in the top half is halfway between half past twelve and a quarter to one, and in the bottom half eighteen minutes to five.

With a two-handed clock there is no need to divide the space between the hour numerals on the inner edge of the chapter ring, since the fine adjustment is indicated by the minute hand (and - even more amazingly to the 18th century clock owner - by the seconds hand when one was fitted). Therefore the *outer* edge of the chapter ring is divided into segments to show the minutes. Figure 3 shows this system in the lower half of the diagram. The diagram does not show the common habit of numbering the minutes at five minute intervals on the extreme outside edge of the chapter ring. Initially the numbers were very small, but as the 18th century progressed the numbers became more prominent in size.

However, for many years it was common for the dial to retain at least one of three concessions to people having difficulty getting used to the new, two-handed clocks. The first was to engrave a mark, often in a fleur-de-lys pattern, halfway between each of the hour numerals (these are shown in the diagram as solid circles); this allowed the old-fashioned customers to continue to use the hour hand to tell the half-hour divisions. The second was to engrave another mark (shown in the diagram as a bold cross), known as a half-quarter marker and often in the form of a lozenge, in the minute band to show the point where the minute hand was halfway between the quarters, thus simulating the old fashioned method of telling the time from the single hour hand. The third was to retain the quarter-hour divisions on the inner edge of the chapter ring. A modern analogy to these practices is to be found in the use of Fahrenheit temperatures in parallel with Celsius for weather forecasts.

In reality, of course, the need for both sets of marks probably disappeared fairly quickly. Nevertheless, both features are to be found on dials up to the middle of the 18th century, and occasionally even later. The fleur-de-lys half hour markers were brought out of retirement for the chapter rings of the so-called Vienna regulators produced in Germany throughout much of the 19th century and into the early 20th century for sale in Britain.

The dial of the Henry Minshull clock (figures 4 and 5) has the main features of both dial styles; in other words it has not only the outside minute band, but also the divisions on the inner side of the chapter ring. This, coupled with the size of the numerals in the minute band (the numerals tended to grow larger as the century progressed), suggests a date of about 1710. Other aspects of the decoration confirm this, and it is pleasing to note the quality of the engraving, including Henry Minshull's signature to left and right of the Roman VI: "Hen. Minshall ... Denbich". Everything about the dial, and the original components of the movement, suggest a competent craftsman producing good quality clocks for the local community.

Considering his obvious proficiency it is a little curious that it was not until 1718 that he appears to have been entrusted with looking after the clock in St. Hilary's. In 1715, for instance, it is a certain John Owen who was paid for fitting "brasse wyre" to the clock and for cleaning it, and in the same year Richard Jones, Smith, was paid one shilling for making a new pulley. In 1718 and 1719 "Mr. Mynshul" ("Mr. Minshul" in the second entry) was paid four shillings and two shillings and sixpence respectively for repairing the clock.

[Figure 4]

[Figure 5 - detail of the signature]

It was probably also Henry Minshull who repaired the clock from Rhuddlan parish church since the churchwardens' accounts note in 1733 that £2 was paid out for "mending the Clok" and a further two shillings "for Cary the Clok to Denbigh and bak again". Two years earlier in 1731 they had paid "the Cloakmaker for vewing y^c Clok" at a cost of one shilling. However, Henry Minshull does not seem to have been involved in 1744 when the churchwardens paid out five shillings and sixpence "for Ale for Eleven Men in Helping to Set ye Dyall plate up".

After his two brief appearances in the Churchwardens' accounts, Henry Minshull did not trouble the annals of Denbigh until his burial in 1753 which is recorded in the brief entry:

> "22th Henry Minshall Clockmaker February 22."

A Mary Minshull had been buried the previous year and, although she is not specifically described as being Henry's wife, it would seem reasonable to assume that this was the case.

The first John Minshull (for there were three John Minshulls in all) was, as we have already seen, a clockmaker by 1722 when he fell foul of the law. The Quarter Sessions order book for 1722 contains the following entry:

> "April 3rd.
> A Bill of Indictment for Petty Larceny having been found ag^t John
> Minshull clockmaker ordered that Robert Simon do take and detain
> him in Custody untill he do find Security himself in 20 l[9] and 2
> Sureties in 10 l Each for his appearance next quarter sess. to traverse
> the s^d Indictment."

The Quarter Sessions roll for the same session then confirms these details and names Henry Minshull and John Shaw as sureties. In the following July, the Quarter Sessions order book takes up the story again with the sentence:

> "July 10
> Ordered that John Minshull clockmaker (who was found guilty of
> petty larceny by verdict of a Jury) be whip^t on Wednsday next att the
> publick cross in Denbigh in markett time and y^t the Goalor of this
> county do take him immediatly into custody, and see that the same
> be done accordingly, and then to discharge him."

The original of this entry is shown in figure 6. The details of his crime are not on record; he may have used the clockmaking business to fence stolen goods or it may

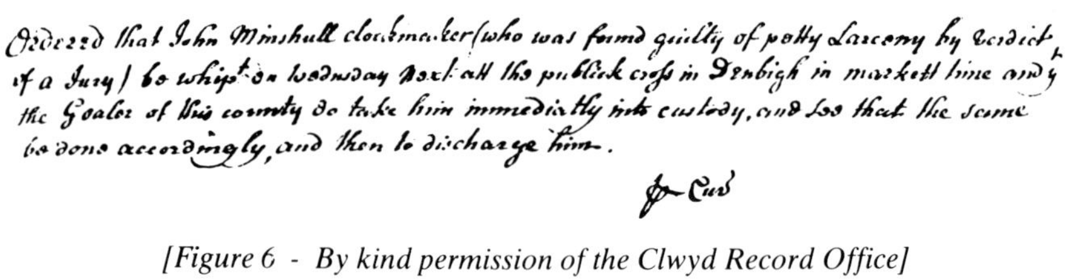

[Figure 6 - By kind permission of the Clwyd Record Office]

[9] 20 l is an abbreviation for 20 librae, i.e. £20.

have had no connection at all with his trade. Either way, it was hardly an auspicious beginning to a career in clockmaking which was to continue until his death over 50 years later.

A glimpse of his family life can be gained through the parish registers. In 1732 we read of the burial of his wife Ann whom he had presumably married in another parish. She was buried on the 20th of July and within two months John had remarried, this time in Denbigh and during one of the Rector's brief Latin spells:

> "September 22 Johannus Minshull Horologicus et Elhora Myddleton
> De Par Denbigh"

They were to have three children. Mary, named after her grandmother, was baptised in 1733 on July 10th, John 6 years later in 1739, and Thomas in 1742, by which time Mary had already died. John was to become a clockmaker in his turn and we can assume that he completed his apprenticeship in or around 1758.

With two John Minshulls working at the same time (the third was to make his appearance in 1786, by when his father and grandfather were both dead) it is often unclear which of the two is under discussion. Occasionally they are distinguished by the adding of Senior or Junior. At other times, however, there is no way of telling whether the father or the son is involved and this may well be an indication of the family nature of the business which they carried on.

I have found no domestic clocks by either John Senior or John Junior, although there is a clock (figure 7), now in Scotland, with a dial signed with the family name, which is further evidence of the close collaboration between the two men. This is a 30-hour clock, that is to say one which must be rewound each day. The stylistic features of this dial point to a probable date of 1760 - 1770 and it is interesting to note that the four corner spandrels are of an unusual design. There are no winding holes, since rewinding is carried out by opening the case door and pulling down the rope.

Inspection of the dial and the quality of its engraving reveals a major discrepancy between the well-executed chapter ring and leaf decoration on the one hand, and on the other hand the cartouche and lettering of the dial signature, "MINSHULL[10]. DENBIGH". This is not uncommon at this period. It was possible, probably even customary, for the clockmaker to buy a fully finished dial to which he would add his own name and town, as required by law. In this case, whoever engraved the Minshulls' name has done so rather clumsily and the effect is cramped; in addition, the engraver has had only limited success in his attempt to echo the leaf decoration by the scrolls added above and below the words inside the cartouche.

A particular quirk of this clock lies in the date wheel, which is clearly visible in the photograph. In most clocks, the hour hand makes two complete revolutions every 24

---

[10] At first sight this looks like "MINSHUIL"; however, on closer examination I believe that the engraver has run the double "L" together. After all, there is a dot over the "I" but not over the first "L"

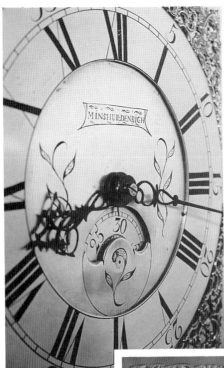

[Figure 7 - Copyright J.D. Waldron, ABIPP, 1992]

[Figure 9 - detail of the signature]

[Figure 8]

hours; thus it is relatively simple to make the date wheel also move forward twice every 24 hours, showing a half-day advance each time. To allow for months of up to 31 days, the date wheel will therefore have 62 teeth on its outside edge, which is hidden behind the dial itself, and its visible face will be numbered up to 31, the owner having to make an adjustment whenever the month is shorter than the full 31 days. (It is of course possible to make a clock which adjusts for the different months and for leap years as well, but this is not found on simple 30-hour clocks, since it would defeat the object of producing a cheap clock). However, the date wheel on this clock has 60 teeth instead of 62 and is numbered up to 30 days. Could the makers have thought that it was perfectly acceptable for the owner to turn the date back by one day at the end of the longest months? Or did the Minshulls just make a simple mistake and then not go to the trouble of making a new, correct date wheel?

Certainly, the other "Minshull" clock which I have seen (figures 8 & 9) has a correctly made date wheel. This is also a 30-hour clock and is also in a replacement case. It has exactly the same spandrel pattern as the Scottish clock, and the spandrels have clearly been made from a worn mould, since the details of the moulding, particularly at both extremities, lack definition. The dial of this clock has no engraving except for the petal-like decoration around the winding holes, which is reflected in the scalloped decoration around the oval cartouche which frames the names "MINSHULL" and "DENBIGH", one above the other. As in the previous clock, the quality of engraving of the names is somewhat clumsy, with the last two letters of "MINSHULL" being considerably smaller than the others and the shape of the letters showing considerable variation in details such as the size of serif. Again, the engraver ran together the last two letters of the surname and remembered to dot his "I"s!

When referring to the Scottish clock I pointed out that 30-hour clocks do not require winding holes. With this second clock, however, which also requires daily rewinding, there appear to be winding holes. Close examination reveals that these are not genuine winding holes. They are too small to take a key and there are no winding squares behind them. They are, in fact, an attempt to make a humble, and cheaper, 30-hour clock look like a proper 8-day clock. This was a common practice and the decoration around the dummy winding holes is an attractive feature of the dial.

The Henry Minshull clock and the two clocks by his son and grandson have lost their original cases. In a sense this is most disappointing, but on reflection it may also be indicative of the quality of Denbigh clock cases. The three clocks would almost certainly originally have been housed in oak cases and the cases would have been constructed by local joiners or cabinet makers. I have already said that Denbigh rectors showed the occupation of the men and women recorded in the parish registers; it is noticeable that there were few cabinet makers in Denbigh in the 18th century and that they all appeared towards the end of the century. The conclusion can therefore been drawn that the cases housing the Minshulls' clocks were probably made by joiners whose methods and materials were not sophisticated.

The Minshulls seem to have been regarded as the leading clockmakers in Denbigh throughout this period, as is testified by the many recorded payments made to them both by St. Hilary's in Denbigh and by the parish council of Bodfari. In the latter case, it is clear that routine oiling and maintenance was carried out by the parish clerk of the day; when a major problem arose, they called in John Minshull:

> "Be it likewise remembered that John Minshull Clockmaker was paid six shillings for mending the Clock y^e 19th day of this Instant and that he is to keep it in repair a whole twelve month for the same money." (August, 1737)

and:

> "P$^{d.}$ for mending the Clock to John Minshull...................0.15.0
> for going to Ruthin for that Clockmaker to have his advice to mend the Clock....0.1.0" (1756)

Why the churchwarden, Robert Jones, should have needed to go to Ruthin to see John Minshull is one of many unanswered questions. A search of Ruthin parish registers reveals no entries relating to the Minshull family but it was perfectly possible for a couple to live in a parish and leave no trace of their presence there; after all, John Minshull and his wife did not disturb the Denbigh parish registers during this period either.

Rhuddlan also had a church clock and they also had problems with it. In the churchwardens' accounts for November 28th, 1762 we read:

> "... it is ordered by us whose names are hereunto subscribed being freeholders Landowners and Inhabitants of the s^d parish that the Vicar & Churchwardens of the said parish are hereby impowered to Contract with a proper Clockmaker (such as they shall think fit) for a Weekly Clock to be set up in the Steeple of the said parish Church of Rhuddlan, and that the Sum of thirty pounds be levied and raised upon the Inhabitants & Landowners of the s$^{d.}$ parish for paying for the said Clock and other Incidents of the parish of ours. As witness our hands etc."

Thirty-one signatures and marks follow this decision, with thirty ayes and only one no, a certain Noah James who placed a grumpy-looking mark beside his name under the heading "Against the Clock". This gave a majority of 29 in favour of the clock and in the following year the parish clerk claimed a shilling for going to Denbigh to see "Mr. Mynshull". We have no proof that the Minshulls actually made the new clock, but the link is yet further proof of their standing in the Vale of Clwyd.

In Denbigh, the Minshulls continued to look after the clock in St. Hilary's Chapel; indeed, they had a contract to "keep the Church Clock of ye s^d Town in good repair" for which they received the annual sum of seven and sixpence. The last year in which this was paid appears to have been 1761, for there is no record of any further payment.

There was good reason for this. The time had come to replace the clock at last and this task was entrusted to the Minshulls. The Churchwardens' accounts for 1765 contain the following contract between the parish and the Minshulls:

> "12 May 1765
> Att a Vestry Meeting held at Saint Hillary's Chappell pursuant to proper notice given, Its Ordered that John M[ll] Senior and John Minshull Junior shall take Down the Clock of the said Chappell and make a good and Substantial 30 hours Clock insted thereof for the sum of four pounds, the said new Clock to be in the same form as the Hall Clock in Denbigh and the said Minshulls to have all the Materials of the Old Clock Towards making a New one, and the said John Minshull Sen[r.] and John Minshull Jun[r] do hereby Engage to make the said Clock in Manner aforesaid and to put the Same up in the said Chappell by the first day of November next and to keep the same in repair for one whole year Gratis, and afterwards to be allowed five shillings p. annum for Cleaning and keeping the same in repair."

This entry bears the signatures of both John Minshulls. In the following year (1766), a Robert Griffiths[11] was paid two guineas for painting the dial and in September John Minshull's bill of four pounds was paid. Subsequently, the Minshulls received regular payments from the parish for their work on the clock, although rarely the five shillings agreed in the contract; usually more (14 shillings in 1768 and over a pound in 1769) but occasionally less, such as the two shillings paid in 1781. By this time John Minshull Senior had died but the bill was still recorded as being paid to John Minshull, just as before.

It is ironic that the Town Hall clock which the Minshulls were asked to copy was the one which Robert Farr had made and it bears out our theory as to how Robert Farr made his clock; after all, the Minshulls were not designing the replacement, but copying a local clock and this despite the fact that they were far more experienced than Robert Farr could possibly have been. It is strange that the Minshulls should have been asked to copy the design of the "Hall Clock", which we know cannot have used the pendulum and anchor escapement. This is borne out by the fact that the Minshulls were able to use parts from the old church clock in the new one. In spite of what we know about the unreliability and waywardness of timekeeping before 1657, Robert Farr's clock was presumably considered a success. It may also be that cost was an important consideration and that the possibility of re-using the parts from the clock which they were replacing made the new clock affordable.

Both the Town Hall clock and the chapel clock have now disappeared, the former to be replaced by a modern turret clock made by Joyce's of Whitchurch (the family connection with the Vale of Clwyd will be explored later in this book), while the new chapel clock was already in disrepair by 1856, since John Williams[12] tells us that "This latter has long since struck its last hour, and is now reduced to a mere skeleton, - a vestige of the ravages of time."

Let us return for a while to the story of the Minshulls themselves. Of the three children born to the first John Minshull and his second wife, Elinor, the eldest, Mary, died before her seventh birthday and the youngest, Thomas, was born in 1742. John II was

---

[11] Robert Griffiths is listed as a clockmaker in both Loomes and Peate and both have spell his name without the final "s"; it seems to me more likely that Robert Griffiths was a painter!

the middle child, being born in 1739, and he was to die at the age of 50 in 1789. Before that, however, he found time to be a clockmaker, get married and father 11 children. His marriage is recorded in the Denbigh parish register on 16th February, 1758, seven months before the birth of their first child, Mary. He signed his name in flowing handwriting; his bride Elizabeth Jones made her mark. The couple produced children at 18 month or two yearly intervals from 1758 right through to 1780, but with an eleven-year fallow period between the birth of John III in or around 1763 and that of the twin boys, Benjamin and Robert in 1774. There is nothing to prove that they were still living in Denbigh during this period, nor is there any proof that they had moved away. Indeed, the exact details of John III's birth are a matter of surmise, as there is no baptismal record in Denbigh or in any of the surrounding parishes. Our only evidence is his recorded age of 82 on his death in 1845. Of the eleven children, only Mary and John III seem to have survived childhood, and even Mary died at the age of 23. The parents chose traditional "Minshull" Christian names for almost all their children; apart from the twins, Benjamin and Robert, all of them were named after parents, grandparents or great-grandparents.

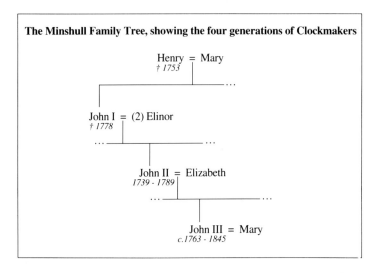

John III married a Mary as his great-grandfather had done, and in 1786 Mary was born, the first of 8 recorded children, again all with familiar names. Indeed, the only new name, Jane, was that of her grandfather's nephew's wife who was married to John Minshull the tanner, and of the 17 Minshull children born in this generation to John the clockmaker, John the tanner and Thomas the shoemaker there were four Marys, four Johns, three Elizabeths and one each of Henry, Anne, Robert, Thomas, Sarah and Jane.

That the third John was a clockmaker is not in doubt, since all references to him describe him as a clockmaker. How good a clockmaker he was is not at all clear. The only clock by him that I have seen was a 30-hour "cottage" clock, that is to say a small long-case clock, and it had a square white dial typical of the turn of the century. In fact, as we shall see in the next chapter, the job of clockmaker had been transformed during the working life of the first two John Minshulls. By the time the last John Minshull was working, the clockmaker's role had changed dramatically and his cottage clock was assembled by him rather than made in the way that his father and grandfather had made the "Minshull" clocks.

There is a gap of nearly forty years during which we hear almost no mention of the last John Minshull, that is to say between the birth of his youngest child, Robert, in 1799 and the working reference to him in Pigot's Directory for 1835, where his address is given as Lenton Pool and the only two occasions on which he is mentioned show that he was having problems.

Documents relating to this period include the Denbigh Vestry Minutes which detail the payments made to the poor and needy. Two of these refer to the last John Minshull. The first reports:

> "Reduc'd John Minshull to nothing p week." (1813)

and the second records:

> "Allowed to John Minshull for to Have One Pound towards His Rent." (1814)

It is clear from these two extracts that John Minshull's problems were of some duration since they had obviously begun before the first entry. It seems that he had been receiving a weekly allowance although the entry recording the decision to start the payment is missing. The second entry appears to be for a one-off payment to tide him over hard times.

Other entries refer to clockmakers' widows: Elizabeth Minshull (widow of John Minshull II and mother of the John Minshull who was in financial difficulties in and around 1813-14) in 1801 and 1812, Elizabeth Sharrock in 1801 and Ann(e) Bartley at regular intervals from 1811 to 1822. These women were all offered financial help in varying degrees.

Whilst many men may have made a successful living out of clockmaking, these entries suggest that the margin between success and failure was not great and that John Minshull III was on the borderline.

Nearly thirty years after these two entries he re-appears in the 1841 census (the first taken in Denbigh) as a watchmaker, still living in Lenton Pool and by now aged 75; this age is inconsistent with his having been born in 1763, but the 1841 census routinely rounded ages up to the nearest five. The entry records that he was born in Denbighshire and married to Mary. Slater's Directory of 1844 records him still working and still in Lenton Pool. The last reference to him is in the parish register for his burial on 8th January, 1845; his age is given as 82 and his address Mount Pleasant.

Unfortunately, while we have been concerned with Denbigh and the story of the Minshulls, we have leapt ahead in time and must now return to the 18th century and to the rival town of Ruthin.

The first Ruthin clockmakers were Edward and William Courter (also spelled Courtier, Couteir and Courten) whose relationship to each other is not known, but since the references to Edward go back to 1724 and those to William go back to 1741 it is possible that they were father and son.

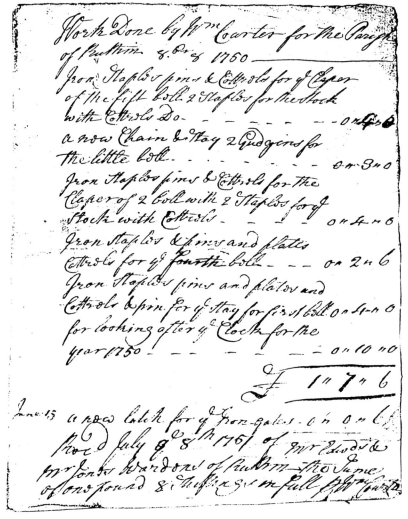

[Figure 10 - By kind permission of the Clwyd Record Office]

The references to Edward are to be found in the baptisms of his children Francis (1724), Sarah (1726), Edward (1731) and a second Edward (1736). A slip of the pen has given him a wife called "Elline" in 1726 as compared with the "Maria" and "Mary" of the three other entries. The last reference is to his burial on June 5th, 1775 and this is the only entry to describe him as a clockmaker. Additional confirmation of his profession comes from Dr. Peate who lists a long-case clock with a brass dial bearing his name, and from an advertisement in *Adam's Weekly Courant*, a Chester news sheet, which describes him as a clockmaker.

Of the two, we know more about William. In 1741 the Ruthin parish register records:

"William Son of William Courtier Smyth & Eliz. was bapt. 8ber 19th"

Two years later William is again described as a smith for the baptism of his son, Edward, on December 17th and again for the baptism in 1746 and burial in 1747 of his daughter Elizabeth. His first wife died, for our next record of him is at his second marriage with Anne Williams, a widow. As with the Denbigh clockmakers, he himself was literate whereas his new wife had to make her mark. William Courter's own signature confirms the correct spelling of his name although the Rector got it wrong yet again. Our last two sightings of William come in 1777 when he married for the third time. His third wife was another widow, Mary Phillips of Llanfwrog. This wedding took place on May 16th. Two days earlier he had been witness at the wedding of Mary Courter to another clockmaker. We shall return to this shortly.

All our references so far have described William as a smith. However, Dr. Peate mentions an article in the *North Wales Weekly News* in 1935 which claims that William Courter "manufactured brass-faced clocks in 1780". How accurate this is must be set against the evidence of the churchwardens' accounts of St. Peter's Parish Church in Ruthin. Figure 10 is a photocopy of one of these, listing the work which William carried out for the parish in 1750. A transcript reads:

"Work Done by Wm Courter for the Parish of Ruthin 8.br 8 1750
Iron Staples pins & Cottrels for ye Claper of the fift bell
2 staples for the Stock with Cottrels Do   0. 4. 0
a new Chain & Stay 2 Gudgins for the little bell   0. 3. 0
Iron Staples pins & Cottrels for the Claper of 2 bell with
2 Staples for ye Stock with Cottrels   0. 4. 0
Iron Staples & pins and plates Cottrels for ye fourth bell 0. 2. 6
Iron Staples pins and plates and Cottrels & pin for ye Stay
for first bell   0. 4. 0
for looking after ye Clock for the year 1750   0. 10. 0
                                               £1. 7. 6
June 15
a new latch for ye Iron Gates   0. 0. 6
Recd July ye 8th 1751 of Mr Edwards & Mr Jones Wardens of Ruthin the Sume of one pound 8 shillings in full Wm Courter"

We have receipts for similar work done in 1747 and 1754. As with Bodfari church, the work of looking after the clock was normally carried out by the parish clerk. Indeed, in St. Peter's churchwardens' accounts we read:

> "Dec. 4th 1729
> ordered that etc etc, ye Clock & Chimes be looked after as usually, y^t ye Clerk be obliged to find wire, oyle, ropes, spades & Mattocks he receiving annually from ye Church-warden ye sum of five pounds."

Moreover, the accounts make reference to a succession of blacksmiths, including Edward Williams (1692), Thomas Roberts (1710 & 1711) and Francis Jones (1710 & 1711), for work done on the clock.

Thus it would seem reasonable to classify William Courter as, at most, an occasional clockmaker, in contrast with Edward who was obviously recognised as a properly qualified clockmaker.

Let us now return to the day when William was witness at the marriage of Mary Courter to another clockmaker. This was in 1777. We cannot be sure what relation Mary was to William or Edward. However, the marriage entry describes her as a widow, and since Edward had died two years before and had been married to a Mary, it is at least possible that she was Edward's widow. The man she married was John Joyce who was related to the famous Joyce family of clockmakers from Whitchurch and whose story will be told in the next chapter.

III

The story of this eminent clockmaking family, who claimed in their catalogue to be "The Oldest Makers of Clocks in the World", has been well told in Douglas J. Elliott's book, "Shropshire Clock and Watch Makers"[13], and a section (pages 153 - 156) is devoted to the Denbigh and Ruthin Joyces. A family tree is also provided which makes tracing the family members considerably easier. Naturally, much of my information comes from Mr. Elliott's research and in the coming pages I shall do little more than try to put the Joyces into the context of the Vale of Clwyd.

John Joyce, who was baptised in 1744, was married four times in all. When he first arrived in the Vale from his native Ellesmere he brought with him to Denbigh his second wife, Elizabeth. They already had two children, Lucy[14] and John from his first marriage. John died and the couple had a son, whom they christened John, and a daughter, Anne. The burial of the second John, is recorded in Denbigh parish register in 1773 and four years later Elizabeth too was buried in Denbigh. Just over three months later, he married Mary Courter and moved with his two daughters from Denbigh to Ruthin.

Mary and John had no children and she died in 1782. This would hardly be surprising if she was indeed Edward Courter's widow, since Edward and Mary had their first

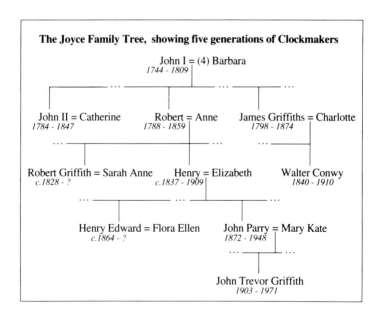

---

[13] published by Phillimore & Co. Ltd., London, 1979
[14] The Joyces often had two or three Christian names and Lucy was baptised Lucy Conwy. For the sake of clarity I shall use only the first Christian name, except where it is helpful to do otherwise. The family tree shows full Christian names.

recorded child in 1724. Assuming Mary to be in her teens in 1724, she would have been in her sixties in 1777; in the same year, John Joyce was 33.

John waited eleven months before his fourth wedding. His new wife was Barbara Griffiths, from Ruthin, and they were to remain married for 26 years and in that time they had John (1784), Peregrina (1786)[15], Robert (1788), Elizabeth (1791), Samuel (1793), James (1798), Thomas (1801) and Sara (1803)[16]. Two of the children, Elizabeth and Sara, died in their infancy but three of the five sons grew up to become clockmakers: John II, Robert and James. When the first John died in 1809, John II took over the family business and for the next thirty odd years, the three brothers plied their trade as far as we know in harmony. Pigot's 1822 directory lists John II working in Welch Street; six years later the 1828 directory lists him, now in partnership with Robert and trading in the re-named Well Street; and from 1835 to at least 1844 the two brothers continued to work side by side in Well Street while the third brother, James, worked in Clwyd Street. John II died in 1847, so that in the 1850 and 1856 directories only James and Robert are listed, and Robert died in turn in 1859. Hardly had one generation begun to fade away, however, than the next grew up to take over the family businesses. John II and his wife Catherine, whom he married in 1828, had no children but between them the other two brothers produced seven sons and one daughter. Two of Robert's sons, Robert Griffith (baptised in 1830)[17] and Henry (baptised in 1837)[18], followed their father's and grandfather's profession and one of James' sons, Walter (baptised in 1840), did the same. Figure 11 shows a watch-case paper advertising Robert G. Joyce's business. It was customary for the maker or repairer to slip one of these discs of paper in the back of the watch case to remind the owner to return to the same firm when further work became necessary. Robert G. Joyce has chosen to decorate his watch-case paper with a fine view of Ruthin Castle and the message that he also dealt in "Gold Rings & Engraving".

As with the Minshulls in Denbigh, the habit of recycling Christian names makes it difficult to follow the family's development and, although it is clear that Robert Griffith continued to trade from his late father's address in Well Street, which later became Upper Well Street, until at least 1889 and that he and his wife Sarah had two children, Robert Griffith (baptised in 1864) and Anne (baptised in 1873), this side of the family disappears without trace at the end of the century.

James Joyce continued to work from Clwyd Street until his death in 1874. In the last trade directory before he died he is listed as James Joyce & Son, the son in question being, of course, Walter who has his own entry in the following directory in 1876 which is printed in bold type and which reads:

"Joyce, Walter C (Watchmaker and Jeweller, and authorized agent to the Singer Sewing Machine Co.) 1 Clwyd Street."

Walter, who did not marry, seems to have built up a particularly successful business. In the 1881 census he is recorded as having an assistant, John Price Williams, aged 22 and

---

[15] Peregrina was named after John's first wife.
[16] Douglas Elliott lists this as 1804.
[17] Mr. Elliott lists his birth as c. 1828.
[18] Mr. Elliott lists his birth as c. 1837

from Efenechtyd, and an apprentice, Edward Davies, aged 18 and from Llanfihangel in Merioneth. A 47 year old servant, Margaret Lucy Whiteside, from Preston looked after them all. Walter Conwy Joyce died in 1910.

The third member of this generation, Robert Joyce's son Henry, is not listed as working in Ruthin in any of the trade directories although by the 1861 census he had set himself up in business in Llanrhydd and had an apprentice, Daniel David Pierce, aged 16 and from Corwen. Daniel David Pierce's older sister, a dressmaker, was staying with them at the time of the census and the family was completed by Henry and Elizabeth's one year old baby daughter Elizabetha and the servant, an unmarried girl of 16 from Derwen. Not long after this the family moved to Denbigh. They are listed in the 1871 census living at 2, Vale Street, by which time there were seven children. R. Conwy, who had been born in Llanrhydd, was 8 whilst Henry Edward, who had been born in Denbigh, was 7. This places the move to Denbigh as 1863 or 1864.

[Figures 11, 12, 13]

If the picture on Henry's watch-case paper (figure 12) is anything to go by, he was at pains to emphasise his loyalty to Denbigh, perhaps because he had moved there from the rival town of Ruthin. The gate-house of Denbigh Castle promotes the image of a solid business and the text describing Henry as "Silversmith. &c." in addition to watch and clockmaker suggests a high degree of expertise.

A brief glimpse into Henry Joyce's private life may be gleaned from a lawsuit in which he was involved and which was reported in the *Denbigh Free Press* in 1883, under the headline: "NEGLECTING TO EMPTY A MIDDEN". The case revolved around whether Henry Joyce had employed Isaac Jones, a labourer, to empty the whole midden in the yard behind his shop for six shillings or to take away six cartloads at a shilling a load. Despite Isaac Jones' protestations, Henry won his case and was awarded four shillings' damages.

At some time between 1871 and 1881, Henry moved to 13 Vale Street and this was to be the family's premises until 1971. Two of Henry's sons became clockmakers: Henry Edward Joyce and John Parry Joyce, and it was John Parry Joyce who carried on the family business at 13 Vale Street until his death in 1948. John Parry Joyce's son John Trevor Griffith Joyce in turn was a clockmaker at the same address until he died in 1971.

The story of the Joyces in the Vale of Clwyd is completed by Edward Henry Joyce who is recorded in Ruthin parish register in 1895 and again in 1897. The first of these two occasions was the baptism of two daughters, Louise and Violet, and the second was the baptism of a son, Henry Conwy Baynton Joyce. On both occasions Edward Henry is described as a watchmaker. The nearest Joyce we have to Edward Henry is Henry Edward, son of Henry, and presumably the two men are in fact one. After all, Henry Edward was correctly recorded in the 1891 Denbigh census as aged 27 but given the name "Henry P. Joyce". It is easy to see how this mistake could have arisen and doubtless there may be an equally simple explanation for the transition from Henry Edward to Edward Henry and then back again. Certainly the watch-case paper (figure 13) shows that he advertised as H.E.Joyce and it also claims that the business had been established for "over two centuries". There is a degree of advertising licence in this statement, but we must assume that he is referring to the whole Joyce family, in which case he was probably tracing the firm back to John William Joyce of Wrexham in 1690.

I said at the beginning of this chapter that we would consider the Joyce family in the context of the Vale of Clwyd and the significance of their clockmaking must be seen in parallel with the success of the Minshulls and of the Griffiths, particularly Richard Griffith whose working life in Denbigh began in about the 1770s. These two families had something of a monopoly over business in Denbigh, and it was the first John Joyce's move to Ruthin to marry Mary Courter and to take over clockmaking where Edward Courter had stopped two years earlier which gave the Joyces the platform to succeed. Perhaps the most surprising aspect of their success was that there was any work at all for other clockmakers in and around Ruthin but from 1822 to 1844 John Parry was listed in the trade directories as working in Clwyd Street, and from 1828 to 1844 David Jones was listed, also in Clwyd Street. Therefore, for nearly a quarter of a century there were five men in business in Clwyd Street and Well Street. It is clear that a small but busy market town such as Ruthin had a high level of clock and watch ownership and that the clocks and watches required time-consuming maintenance and repair.

In order to tell the story of a single family it has again been impossible to avoid bounding ahead into the 20th century, which is not strictly speaking inside the intended scope of this book, so we shall now behave sensibly and return for the last time to the 18th century to consider some less prominent clockmakers.

IV

So far, we have considered clockmakers who established themselves, their families and their businesses in the Vale of Clwyd over a substantial period of time.

However, as I have already said, it was possible for a man to live in the 18th century and leave very little trace of his existence. Hugh Edwards was one such man. In 1735 the churchwardens of Ruthin recorded paying him thirty shillings "as was allowed for looking after ye Clock" and in 1767 the Rector of St. Peter's in Ruthin recorded his burial:

> "Hugh Edwards Clockmaker 27 Feb^{ry}."

This is the sum total of our knowledge of Hugh Edwards.

Fortunately, the majority of men married and proceeded to produce children. Equally fortunately, the rectors of Denbigh and Ruthin usually recorded the men's professions. As a result, we have a reasonably clear picture of the clockmakers in these two towns during the 18th century.

For four of them, there is only a single entry in the parish registers:

> "Owen Williams (Watch Maker) was bury^d July ye fifth." (Ruthin, 1756)

> "Elizabeth Wife of Thomas Garnett Watchmaker, buried 26th January." (Denbigh, 1775)

> "Samuel, Son of Snead Read Watchmaker, By Jane his Wife, was born October the Seventh, Baptized the Twenty Sixth of October." (Denbigh, 1793)

> "Hugh, Son of Hugh Williams, Clockmaker, by Elizabeth his wife, was born the Twenty fourth day of May, Baptized the Second day of June." (Denbigh, 1794)

These entries are of course not conclusive proof that any of them lived and worked in Ruthin or Denbigh and, as far as the last two are concerned, even if they did, their stay was a fleeting one. Presumably they came to Denbigh with the intention of settling but then found that there was not enough work to go round, as must have been the case with another man, David Davies, who certainly spent at least two years in Denbigh, long enough for Susana [sic] and Anne to be born in 1738 and 1740 respectively and for Anne to die at the end of the same year.

Another man, William Somner, is recorded twice in Ruthin parish registers but the entries cover only the time from September 6th 1776, when his daughter Elizabeth was baptised, to September 19th of the same year, when she was buried.

Even after eliminating these clock- or watchmakers from our enquiries, there still remain four who worked in Denbigh for more than the odd year or two.

The first of these is David Foulkes. He is yet another excellent example of how a man could live and leave very little documentary evidence. The only references to him which we have are a footnote in John Williams' "Ancient and Modern Denbigh"[19], his will dating from 1732 and his burial record in 1736. In John Williams' book, he is mentioned in the list of contributors to the establishment of the Free Grammar School, his share of the £339 12s. being £5. His burial entry reads:

"David Foulkes of Henllan Street watchmaker was buried May 25th."

(The entry is in Henllan parish register; one side of Henllan Street lay in the parish of Denbigh, the other in Henllan). His will is the most complete document available to us but throws little light on his work. Nevertheless, it does show that he was unmarried and it confirms that he lived in the borough of Denbigh but in the parish of Henllan. His whole estate was bequeathed to nephews and nieces and their children and included two bequests of ten pounds, his house and gardens and "All the rest and residue of my Goods Chattells Bills Bonds and all other my personall Estate whatsoever and of what nature soever the same Consists". His two nephews, Edward Foulkes and David Roberts, were to receive "all my Working tools belonging to my Trade Share & Share alike." As far as we can tell, neither of these men was a clockmaker and presumably the tools found their way into the hands of a local clockmaker.

David Foulkes, then, was successful enough to need to make a will, literate enough to sign it himself and sufficiently well off to be able to make gifts of £5 towards a new school; he was sympathetic to the educational needs of the town at that time; and he was a genuine working watchmaker with all the workshop equipment to carry out his trade. Unfortunately, however, in the absence of any examples of his work, it is impossible to answer the three most interesting questions of all: namely, how long he worked in Denbigh, what sort of watches the people of Denbigh were willing to buy, and how good a craftsman he was.

After David Foulkes, we come to Christopher Sharrock. He lived in Denbigh from before 1759 until his death in 1783. We know that he was already settled in the town before 1759 because his marriage entry reads:

"19th May 1759
Christopher Sharrock & Elizabeth Jones both of the parish of Denbigh."

He signed his name and, like so many other clockmakers' wives, Elizabeth made her mark. A son, Thomas, was born in the following year; three years later a daughter, Margaret, was born and died; and in 1782 the parish register makes the following entry:

"Christopher Sharrack Son of Christopher Sharrack by Elizth his wife formerly Elizabeth Jones Aged 17 Died Sepr 8th Buried Sepr 10th."

Perhaps the rector overlooked the birth in 1765 or it may be that the boy was not baptised. Christopher Sharrock himself died in 1783 and Elizabeth outlived him by 21 years, dying in 1804. It appears that Christopher had married a younger woman, since

[19] Op. cit., p. 297

his recorded age at death is 71, whereas hers is 79. Simple arithmetic will show that, assuming these ages to be correct, he was 47 and she was 34 when they married in 1759.

Unlike David Foulkes, Christopher Sharrock has left us one example of his work: a long-case clock (figures 14, 15 & 16) living only a few miles outside Denbigh. It is housed in a dark oak case, heavily carved and quite possibly not original. Across the whole width of the hood above the dial is the message "use . time . wisely". It may be instructive to compare the dial of this clock with the two "Minshull" dials. The stylistic similarities are clear at a glance: the date aperture is of the same type, the corner spandrels are made up of rococo S and C shapes (although the exact patterns are not identical), the relative sizes of the Roman hour numerals and Arabic minute numerals are very close. There are differences too: the matted centre of the Sharrock dial has no engraved decoration, the winding holes and seconds dial on the Sharrock clock show that it is an 8-day movement as opposed to the less expensive 30-hour movement on the Minshull clocks. Nevertheless, it would be safe to date Christopher Sharrock's clock as belonging to the same decade as the Minshulls' clocks, that is to say the 1760s.

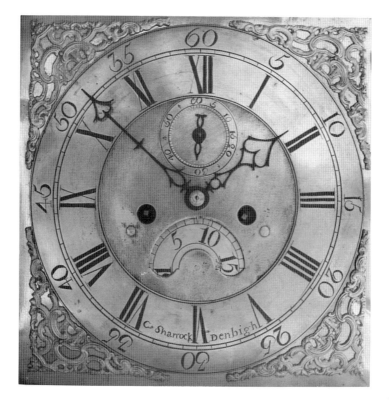

[Figure 14]

Despite the overall similarities between the two clocks, there is one glaring difference. The Minshulls, on the evidence of the dial we have examined, were not proficient engravers; they therefore bought engraved dials from a specialist engraver, probably from no further afield than Chester or Wrexham. Christopher Sharrock was no better an engraver than the Minshulls, but he opted for a cheaper solution, that of engraving the whole dial rather than just his own name. His uneven signature with its spidery star of a full stop, the uneven edges to the Roman hour numerals, the wavery Arabic minute numerals, the 35 where he should have engraved 55 at eleven o'clock and the inconsistent numbering of the seconds ring which uses 5 and 15 but switches to arrow symbols for 25, 35, 45 and 55, are all features pointing to a clockmaker doing his own engraving and not doing it very well. Moreover, he has had difficulties with the top of the figure 1 in both the 10 and the 15 on the seconds ring. The quality of his engraving is all the more evident, because it can be compared with the engraving of the date wheel which is crisp and professional.

[Figure 15 - detail of the signature]

[Figure 16 - detail of the seconds ring]

It may be, of course, that I am doing Christopher Sharrock an injustice. He too may have bought in the dial but have chosen a less skillful supplier than his competitors, the Minshulls. However, it is far more likely that he engraved the dial himself, and this clock is therefore an excellent example of the standard of work which could be expected from a local clockmaker. Moreover, let us add that these idiosyncracies considerably enhance this clock's charm and that the maker's decision to do his own engraving must have made some economic sense. In general terms, 8-day clocks cost about twice as much as 30-hour clocks throughout this century; by engraving his dial himself, instead of having to buy in a more expensive ready-engraved dial, Christopher Sharrock may have been able to keep a competitive edge to his prices.

It would be interesting to know whether he engraved all his own dials. After all, he must have been in his forties and perhaps even his early fifties when he made this clock and the clumsy engraving could indicate that this was an early attempt at engraving a complete dial, implying that his previous clocks had used professionally engraved dials. On the other hand, since he was well into middle age by the time he made this clock, it is possible that his manual dexterity was beginning to deteriorate. A third possibility is of course that he had always done all his own engraving and that he had just never got the hang of it.

It is intriguing to speculate on the customers' reaction to the imperfections on this dial. Did they notice the inconsistencies and mistakes, in particular the 35 at eleven o'clock, and if so, did they ask for and perhaps receive a discount? Is it possible that they noticed nothing amiss and that Christopher Sharrock felt it inappropriate to point out the blemishes? If this was the case, did any visitors to the house spot the inaccuracies and, if they did, would they have regarded it as too delicate a matter to mention to the clock's proud owners?

Before we move on, let us remember that Christopher Sharrock belonged to that generation of clockmakers who made their own clocks and this is therefore an ideal opportunity to discuss the transformation in clockmaking which occurred at the end of his life and to which I have referred earlier.

The revolution in the way clocks were made in the provinces coincided with the appearance of white painted dials, although the two trends happened independently of each other. Until about 1770, local clockmakers had to make all their own working parts for their clocks. Dials and cases could be bought in, but the mechanical components had to be made. Using basic machinery and the power of their own hands or feet, working in natural light or poor artificial light, the clockmakers took raw brass or iron bars, rods and sheets and turned them into accurate and hard-wearing working parts. We may be a little critical of their engraving skills, but their products are still working as accurately as ever after a working life of well over two hundred years.

By the 1780s, the provincial clockmaker could do what his London colleagues had done for many years, that is to say, he could buy in ready-made components from specialist suppliers. Instead of having to make his own gear wheels or escapement pallets, he could buy them in and assemble them. This must have been an economically viable way of producing clocks because, as soon as it became possible, everyone began to do it. Indeed, it was soon possible to buy complete movements, which the clockmaker installed in a ready-made case behind a ready-made dial which the manufacturer would very obligingly have "signed" with the local clockmaker's name. The clockmaker's job had been transformed, but he still gained all the publicity.

The 18th century in Denbigh ended with the appearance of a new clockmaking family, the Griffiths, who made excellent use of the new working practices, whose influence is still alive today and whose work will be the subject of our next chapter.

Richard Griffith died in 1835, aged 80. There is a record in Henllan Parish Register of the baptism of a Richard,

> "Son of Luke Griffith of Bannister Isaf by Elizabeth his Wife"

on March 16th, 1755. The date is about right but we have as yet no proof that this definitely refers to our man. At all events, Richard was already nearly 40 when he first appears in the Denbigh parish registers on the occasion of his marriage on March 8th, 1793 to Mary, née Edwards, whose father was an innkeeper in Abergele. They were to have eleven children in the twenty years between 1794 and 1814. Of these eleven, two became clockmakers: David, who was born in 1800, and their youngest child Richard, who was born in 1814. Richard died in 1841 aged only 26 but David, as we shall see, was a clock and watchmaker throughout his long life, fathered two clockmakers (David Lloyd and a third Richard) and became famous as Clwydfardd, the first Archdruid of Wales.

There is a description of Richard by William Williams, a schoolfriend of one of Richard's sons, which can be translated thus:

> "Mr. Griffith was a loyal member of the Wesleyans for 33 years. He was greatly respected by his fellow men, was very knowledgeable with a strong memory, interesting and humorous, and was a strong supporter of all branches of Welsh literature."

Of the Denbigh clocks that have survived, there are more by Richard Griffith than by any other maker. The earliest (figures 17, 18 & 19) has a brass dial and is typical of the 1780s. It must have been a fairly expensive clock since not only is it housed in a handsome mahogany case with satinwood banding and inlay and a fine hood but it also has a moon dial in the hood arch. The least satisfactory aspect of the clock lies in the dial signature (yet again!); however, Richard Griffith made a bold attempt to decorate his signature, with leaves sprouting from the tips of the letters and elongated tails to the "G", the first "f" and the final "h", although for some reason he omitted his initial. In 1872 the clock must have been repaired by the firm of James Joyce & Son in Ruthin, for the following verse is written inside the case door in pencil:

> "Here I labour with all my might
> To tell you the hour of day and night
> Therefore example take from me
> And serve thy God as I serve thee."

together with the name of the firm and the date 6 / 3 / 72.

Also dating from about 1785 is another 8-day long-case clock, this time with a painted dial and signed R. Griffith. It is very similar to the previous clock in that the date aperture is of the same pattern and the individual minutes are shown by a series of dots, as are the seconds on the seconds ring and the days on the date wheel. The hands on the two clocks are of totally different designs, but hands are easily damaged and easily replaced. The second clock has a more modest case, made of oak with a small amount of mahogany cross-banding.

These two clocks from the same period probably represent his middle of the range oak-cased product and his most expensive mahogany-cased clock; bottom of the range would still have been an oak-cased 30-hour clock. Clockmakers were not high-octane scientists but business-men and the clocks they sold were in response to public taste. During the mid-1770s the white-painted dial clocks replaced brass-dial clocks almost overnight and in the years that followed Richard Griffith would have made fewer and fewer brass-dial clocks and more and more with the fashionable white-painted dials. It should be noted that the new dials were not cheaper than the brass dials they ousted; in fact if anything they were slightly more expensive, but they were so legible and so much more colourful that people were willing to pay the extra money for the extra benefits.

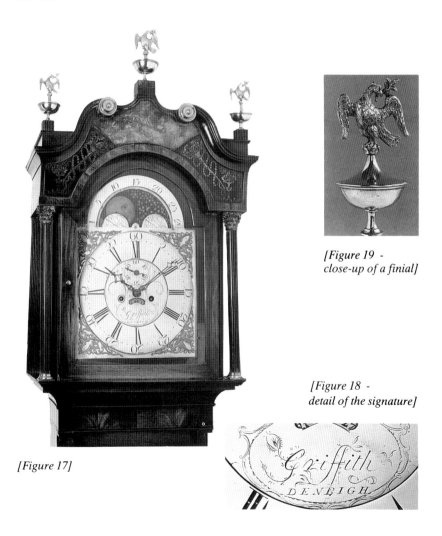

[Figure 19 - close-up of a finial]

[Figure 18 - detail of the signature]

[Figure 17]

This must have been a relatively profitable time for clockmakers throughout the British Isles. Clocks were in demand and they were easier to "make" than ever before. If proof were needed, it can be seen in a movement and dial (the case is sadly damaged) which date from about 1815. The dial is signed "R. Griffiths, Denbigh", but the bell and the backplate of the movement are both stamped with the name "G. AINSWORTH - WARRINGTON". (George Ainsworth was a well-known bell-founder and supplier of materials around 1810 to 1815). The clockmaker's involvement in producing this clock was to order and assemble the parts, to fit the dial which came from the factory of Walker & Finnemore[20] in Birmingham, and then to adjust and regulate the clock before fitting it into the case which a local cabinet maker would have produced. Finally, it was delivered to the customer and set up in his home.

This is not to say that clockmakers had suddenly become gentlemen of leisure. They were still just as busy as ever, but they gradually ceased to build clocks and began to concentrate on assembling and servicing them. The servicing included not only regular cleaning and oiling, for oils were no more effective than they had been at the beginning of the 17th century, but often also a contract to wind and adjust the clock every week. By means of a weekly visit, the clockmaker would be alerted to any problems that were beginning to develop.

[Figure 20]

[Figure 21]

Yet another of Richard Griffith's long-case clocks (figure 20) is in Denbigh itself. It also dates from about 1815 and is typical of its time. It has a pretty, white-painted

[20] Walker & Finnemore are listed as working in Birmingham from 1808 to 1811.

square dial and is housed in a well-proportioned oak case with some banding. The hands do not match and the brass finials look unhappy, but the clock is otherwise in very good condition and has obviously been well looked after for many years. Other long-case clocks by Richard have found their way to Caerwys, to Mold and to Washington D.C. Judging by the number of his clocks that have survived he was the most successful maker and salesman of his generation not just in Denbigh but in the Vale of Clwyd.

Richard Griffith also sold watches, two of which are in the collection of the Welsh Folk Museum in St. Fagan's. Both watches are pair-cased, that is to say they have a second, outer case to protect the movement from dust and the climate; both are hall-marked for Birmingham, 1825 and bear the same silversmith's mark, JH. They both use verge movements with fusee (a device which evens out the otherwise uneven force of the spring) and the backplates are finely engraved, signed and numbered 510 and 515 respectively. Watch n°. 510 is shown in figure 21. The quality of engraving on the centre of the dial is also good and both watches have the same rhyme engraved in capital letters around the outside edge of the bezel:

"KEEP ME CLEAN AND USE ME WELL
AND I TO YOU THE TRUTH WILL TELL"

It appears, then, that Richard Griffith bought his watches from a Birmingham factory. He was certainly buying to a price because, although these watches are nicely finished, they are hardly distinguishable from any of millions of other watches manufactured during this period. In fact, they were about to become obsolete since the lever escapement was already being developed and would be produced in increasing numbers from about 1830 onwards; but Richard Griffith cannot be blamed for selling his customers good, solid, reliable watches of the type he had been promoting throughout his working life.

During the second half of Richard Griffith's working life, trade directories for North Wales began to spread and tradesmen in the vale of Clwyd had new opportunities for advertisement. The first such directory, Holden's, appeared in 1811 and there were to be 16 directories by the end of the century, published at regular intervals. They are a useful source of information about the names, addresses and allied activities of clockmakers, but need also to be treated with caution as the information they give can be highly unreliable. This is the case with Richard Griffith, whose name is recorded in four directories (1822, 1830, 1835 and 1840). It is useful to discover that his trading address is given as Vale Street but confusing to read that Richard Griffith is still working in 1840, five years after his death. There are other posthumous entries for Denbigh clockmakers; indeed, it will happen to both John Williamses! On this occasion, however, there is a logical explanation.

Let us remember Richard's son, Richard, who was born in 1814 and whose death is recorded in 1841: "Richard Griffith; Watchmaker; High Street; Age 26". Although we might have expected the 1835 directory entry to add an "& Son", since by then Richard junior should have been making a full contribution to the family business, we can assume that the Richard of 1840 is the son who was to die the following year. In addition to his listing as watchmaker, the Richard Griffith of 1840 also has entries as:

"Griffith Richard Tin-plate work & Braz'·"

and:

"Griffith Richard Watch and Clockmaker, Ironmonger and Agent for the Atlas Fire and Life Office."

This Richard seems to have been an ambitious and energetic young man with a penchant for diversification. The same is true for his elder brother, David.

[Figure 22 - By kind permission of the National Library of Wales]

David (figure 22) was born in 1800. At the age of 11 he left school to learn his father's trade but, unlike Richard, he then left home. This was in 1828. In 1831 he was in Caernarfon where he stayed until 1842; by this time he was married to Eleanor, née Jones, from Barmouth whose father was a sea captain engaged in coastal and Irish trade. David and Eleanor had three sons and five daughters, all of whom reached the age of thirty plus. The second child was their first son, whom they baptised Richard; from the information in the 1881 census this must have been in 1839 or 1840. They then spent five years in Amlwch before returning to Caernarfon, where David Lloyd was born in 1849, for a further two years and subsequently moving to Holyhead. Finally, in 1856 they returned to Denbigh and remained in business there until 1881. The following year Eleanor died and David moved with his daughter to live in Abergele, where Eleanor's family had come from.

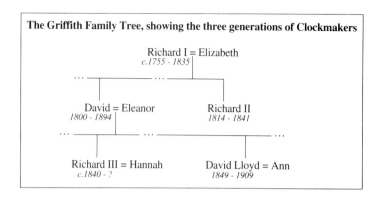

David is best known for his role in helping to shape the modern Eisteddfod and doubtless his father's strong interest in Welsh literature, to which we referred earlier, was crucial in awakening David's own passion for the Welsh culture. This is not the primary subject of our interest but it may be useful to reflect that his success in the cultural arena mirrors the status of the clock-maker in his society; clockmakers were relatively well-educated and their trade enjoyed the connotations of scientific progress which we might associate with electronics today. It is no surprise that men such as David Griffith and, a few years later and in a more restricted sphere, Wilhelm Keepfer had the qualities necessary to make their mark in the public domain.

As we have seen, David returned with his family to Denbigh in 1856 and in the first directory after this (Slater's of 1868) Griffith & Son are listed in Vale Street. At this time Richard was probably living in St. Asaph, where the 1861 census shows him aged 21, living in the High Street with his sister Sarah, aged 15, acting as his housekeeper. Our next reference to the family comes in the 1871 census when we find David and the as yet unmarried David Lloyd at 8 Beacon's Hill while Richard, who has married Hannah from Liverpool and his junior by ten years, is living at 9 Vale Street. The next two directories (1874 and 1876) both show Griffith & Son trading from 9 Vale Street, but by the end of 1881 the three men had gone their separate ways. David had retired to Abergele, where he continued to work and where his grandchildren remember him overhauling the church clock; Richard was trading at 36 High Street, Rhyl, where he employed one man; and David Lloyd had stayed in Denbigh where his premises in the 1881 census and the next two directories are given as 3 Vale Street. On the evidence of the 1886 directory he had followed the family fondness for branching out because he is described as "watchmaker ironmonger and general home furnisher". Like his uncle, the second Richard Griffith, he also dabbled in insurance, being the agent for the Scottish Widows & Atlas Insurance Offices, a sideline which he ran from his home at 37 Vale Street.

A watch-case paper which Griffith & Son used is shown in figure 23. The design suggests a solidly traditional business, with the use of a royal crest, while at the same time reminding us that the Griffiths were Welsh. This is the only watch-case paper from the Vale of Clwyd to make any use of the Welsh language, even though Henry Joyce, just to take one example, was bilingual.

Our last but one sighting of David Lloyd is in Slater's 1895 directory. He is now living in 5 Vale Street and is described as "watchmaker, ironmonger and coach proprietor". (His name is given as Lloyd, David but there is no doubt that this is no more than typical directorial muddle). He died in 1909 and the *Denbigh Free Press* printed the following report:

> "The death of Mr. David Griffiths took place at 3.30 on Monday morning at his residence in Vale Street. The deceased had been in business in Denbigh for many years as a watchmaker, and was well-known throughout the Vale of Clwyd. For some time he had been unwell, and under medical care ... The deceased's father was a well-honoured Welsh bard of great renown, being Clwydfardd, the Archdruid. Deceased leaves a wife and a number of children."

We are fortunate in having a description of David Lloyd Griffith among the reminiscences of a contributor to the *Free Press*, who wrote under the pseudonym of "Dyn o'r Stryd" (Man of the Street), describing trade and shopkeepers in Denbigh during the 1880s and 1890s. I shall quote what he has to say of David Lloyd Griffith in an English translation,

> "I remember him first keeping the present shop of W.H.Smith [now Cronin's]. There was one window full of clocks and watches as well as jewellery and the other full of new and second-hand furniture and all sorts of household ironmongery. Inside the large shop there were various clocks, not displayed very tidily. Occasionally he held furniture sales in the fowl market opposite. This was the time Dafydd Griffith was seen at his best. He was a very loquacious fellow. If the first bid he received for a piece of furniture was pitifully low, he would tell him to keep his money to pay his weekly contribution to the 'funeral club'[21]. Everyone enjoyed his wit and humour. Like his next-door neighbour, Harry Davies, he was original and outstanding among Denbigh shopkeepers."

Several clocks survive from the second half of the century, two of them in chapels and a third in a private home. The chapel clocks, which are both wall clocks, are signed Griffith & Son (presumably dating from the period 1856 to 1881) and David Lloyd Griffith (1881 to 1899) and the domestic long-case (figure 24) is signed Griffith. Or it would be truer to say that it was originally signed Griffith, for repeated cleaning of the dial has entirely erased the signature. This clock is entirely in keeping with the fashion of 1870 and thereabouts and it may well have been given to the present owners' grandparents as a wedding present in the 1870s. The case is broad and solid and the dial decoration has lost the freshness of the dials of the first half of the century. This clock, in fact, typifies the end of the long-case clock's lengthy career. In London, long-case clocks ceased to be fashionable at the end of the 18th century but they continued to be

---

[21] It may be necessary to explain to some readers that money was paid into the 'funeral club' to pay in advance for one's funeral costs.

popular throughout the rest of the British Isles for a further 70 or 80 years. By 1900, though, the long-case had gone out of fashion and had been replaced by a variety of much less expensive types such as French black "marble" clocks, American clocks in all shapes and sizes and the popular German wall "regulators".

The history of clockmaking in the Griffith family, spanning as it did the years from about 1775 to 1900, witnessed all of this; indeed, the three generations went from the age of genuine clockmakers right through to the beginning of the collapse of the British clockmaking industry.

[Figure 23] *[Figure 24]*

VI

It has probably become clear in the course of the previous chapter that much more information is available about the 19th than the 18th century. Our sources are widened to include trade directories (although we have seen that these can be unreliable), newspapers and census returns. The first census in Wales to record personal details was in 1841 and the number of details recorded grew with each new census. Thus, in 1841, the place of birth is not specified; the census officer merely indicated whether each adult was born in Denbighshire (Flintshire for the parishes of St. Asaph, Rhuddlan and Rhyl) or not. However, by 1891, information had become far more precise and included details on bi-lingualism. The main gap lies in the 1861 Denbigh census which is incomplete and includes no clockmakers at all.

Denbigh had a full complement of clockmakers all through the 1800s. During the 18th century there never seem to have been more than four working clockmakers at any one time. In 1810, by contrast, there were six, in 1880 there were also six and in the final decade there were no fewer than eight. A similar story is repeated in Ruthin and, of course, the number of jewellers and clockmakers in Rhyl mushroomed in keeping with the town's rapid expansion.

As in the previous century there are a number of makers in the Vale for whom we have only one reference. For example, the adventurously named Gustavus Adolphus Wilson married Eleanor Jones in Denbigh in 1815 when he is described as "Batchelor, Watchmaker of this Parish"; he probably moved away to live in Liverpool, since a Gustavus Wilson is listed by Baillie[22] as working there in 1825. Edward Greatrix showed that watchmakers are all too human in the following entry in Denbigh parish register:

> "February 26 Edward Greatrix base born son of Edward Greatrix
> Vale St Watchmaker and Elizabeth Hughes Vale St."

In St. Asaph, Robert George Kelly's presence is registered at the baptism of his daughter, Florence Elizabeth in 1894. In 1851 we find that James Perch, from Peckham in Kent, has set up as a Clock Maker in Ruthin with his wife and four children and in 1871 Selina Scanlan, although she is only 16, is described as a Watch Maker at her mother's address, also in Ruthin. In the 1891 Rhyl census there are no fewer than five clockmakers or watchmakers for whom we have no other reference and in all a total of 47 clockmakers troubled the registers, directories and censuses only once during the 19th century.

Following the mention of Selina Scanlan, this is perhaps also the place to record that the Vale of Clwyd appears to have produced only four women watchmakers and that only one of them is recorded more than once, suggesting that the Vale was not a hospitable area for them. In addition to Selina Scanlan in Ruthin we have Annie Louise Harris and Eleanor Jones both in Rhyl, and Elizabeth Rigby who is described as a jeweller in the 1891 census in Ruthin but advertised in the clock and watch makers section in the 1889 directory. Elizabeth Rigby and her family established themselves for a while as a successful business, with branches in Bala and Ruabon, and a shop is recorded in the 1895 Denbigh directory as Rigby & Rigby, trading in Bridge Street.

---

[22] "Watchmakers & Clockmakers of the World, Volume 1", 3rd Edition, N.A.G.Press, 1989; p. 345

According to the 1891 census, Elizabeth was a widow and her cousin William, who was born in Ashton-in-Makefield, worked as her manager. Elizabeth's two teenage daughters and a nephew, who is rather ambiguously described as a "Mechanical Student", lived with them in their premises at 14 Market Place. Another relative, Thomas Henry Rigby, and his wife, Esther, also settled in Ruthin for a short spell and just long enough for their daughter, Marie Vernon, to be born there in 1900.

Among the single references are those for a series of apprentices. John Davies (1861) was employed as apprentice in Rhyl by William Jones. Richard Evans (1871) and Henry Edward Joyce (1881) both learned their trade under Henry Joyce, while Phillip Meisinger (1871) and Jacob Bauer (1881) were apprenticed to Wilhelm Keepfer. As we saw earlier, Daniel David Pierce was apprenticed to Henry Joyce in Ruthin in 1861 and in 1881 Edward Davies was apprenticed to Walter C. Joyce.

The majority of these men disappear from the Vale without trace and their number suggests that it was not easy to establish a business in 19th century North Wales. It is clear from the census information that many of these clockmakers had come from far afield and, as we shall see, relatively few of the clockmakers of Rhyl were born locally. For a few, it is possible to follow at least part of their lives. We already saw Gustavus Adolphus Wilson moving to Liverpool, and Henry Edward Joyce, who was apprenticed to his father, went on to become a clockmaker in his own right. Given the distinctive combination of names, it seems reasonable to suggest that Henry Joyce's apprentice, Daniel David Pierce, moved on to Wrexham and Ruabon as recorded in Dr. Peate's list[23]. It seems equally reasonable to suggest that Edward Griffiths, listed in the 1861 census in Rhyl, was trading in Bangor in 1844 (the census shows that Edward's son, Edward M., was born in Bangor in 1845) and went on to work in Abergele where he is listed by Dr. Peate[24] from 1868 to 1893. Also reasonable is the assumption that Robert George Kelly, who was working in St. Asaph in 1894, is the Robert George Kelly who was recorded in the 1871 Denbigh census as David Price Jones' stepson, then aged 3 and a "scholar", and in the 1891 Denbigh census as David Price Jones' son, then aged 24, already a widower and a "Jeweller's Assistant". (Since 1871 David Price Jones had abandoned his appellation of "Watchmaker" in favour of "Watchmaker (Master)" in 1881 and "Jeweller (Master)" in 1891). Robert George Kelly appears to have spent some time in Llangollen because his son, aged 15 months in 1891, is recorded as having been born in that town.

The one set of names which is hard to trace definitively is the William Joneses who traded in the Vale in the mid 19th century. Four[25] are recorded, two in St. Asaph, and one each in Denbigh and Rhyl. In reality, there may have been only two men, separated thus: One man was working in St. Asaph in 1850 according to Slater's directory. This man could have moved to Rhyl where he worked from 1851 to at least 1874. The second man was also working in St. Asaph in 1850 (Slater's directory again), but stayed longer, being recorded in the 1851 census and at the very end of 1851 at the baptism of his daughter, Harriet. These dates would allow him to have moved to Denbigh where Slater's 1856 directory records a William Jones as working in Vale Street.

[23] Op. cit., p. 67.
[24] Op. cit., p. 44.
[25] A fifth William Jones is recorded in Dr. Peate's list as working in Ruthin, but no dates are given.

This second William Jones was born in Llanrwst, several miles to the West of the Vale of Clwyd and home to the famous Owen family of clockmakers, and married Elizabeth who came from Ruabon, not far from Wrexham. The move to St. Asaph, and perhaps on to Denbigh, was therefore not a huge step for either of them but he and his family had made the journey from Llanrwst to St. Asaph by a circuitous route. The couple's three oldest children were born in Manchester between 1838 and 1843 and their three youngest children were born in Liverpool between 1845 and 1849. All this information is recorded in the 1851 St. Asaph census.

The three Thomas Bartleys are at first sight equally hard to pin down and I must admit that for some time I thought that the rector of Denbigh had made a silly mistake in the burial register. After all, it has to be surprising that two Thomas Bartleys should die within a few days of each other, one on the 14th and the second on the 19th of August 1811. Moreover, they could hardly be, for example, father and son, since their ages are given as 30 and 39 respectively. My first suspicions were reinforced, although only slightly, by the discovery of a single gravestone in St. Marcella's churchyard :

> "Here lieth the body of Thos Bartley Clockmaker who died Aug. 19th 1811 aged 39"

whereas Dr. Peate affirms that both men are buried in the churchyard[26]. Soon, however, I realised that I had unfairly criticised the rector, for further research showed that there had indeed been two different men with the same name and the same trade. One of them came from Pwllheli and married Anne[27] (née Parry) in 1798; they had a son, John, in 1809 and a daughter, Hannah, in 1811, whose baptisms were both recorded in the Swan Lane Chapel register. It is presumably Hannah's burial which was recorded in the following incomplete entry in the parish register:

> "_____ daughter of the late Thomas Bartley Clock^r by _____ his wife, formerly _____ Parry died Octr 4th buri^d 7 an Infant"

All we know about the other Thomas Bartley is that he was married to Jane (also née Parry) and that they had a son, John, who died in 1808.

A Thomas Bartley is listed in Holden's 1811 Directory and again in their 1816 Directory, by which time both Thomas Bartleys had been dead for five years; but the Thomas Bartley affair does not quite end here. A third Thomas Bartley came to live and work in Denbigh. He was born in 1799 or 1800 in Pwllheli, which suggests a family connection with the first Thomas Bartley who was also from Pwllheli. He is recorded in both the 1851 and 1871 censuses (remember that the 1861 Denbigh census is incomplete). In 1851 he lived in Lenton Pool and was married to Mary; by 1871 his wife had died and he had moved to 108 Henllan Street, where he was a lodger..

It is again frustrating to have to admit that no clocks by any of these three Thomas Bartleys have appeared[28], even though between them they must have worked for a total of upwards of 25 or 30 years.

[26] "Clock and Watch Makers in Wales", Dr. I.C.Peate; Cardiff 1960; p. 32
[27] This is the Anne referred to in chapter II, page 21 , as receiving financial help between 1811 and 1822; she was still alive in 1851, when the census describes her as "Clockmaker's Widow".

Almost as confusing as the John Minshulls and the Thomas Bartleys are the Williams families. There were three John Williamses, none of them related to any of the others. The first, often referred to as John Williams Senior, was born in about 1770 outside Denbighshire and was working in Denbigh by 1802 when a son, Robert, was born to him and his wife, Mary. In all, they had five children, three boys and two girls. From the addresses in the parish registers, the census returns and the trade directories, it would seem that he lived in Swan Lane but that his business was run from the High Street (1830 and 1835) and then Hall Square (1840). John Senior died in 1842 at the age of 72, or 71 if you prefer to believe the inscription on his gravestone in St. Marcella's churchyard. Two years after his death he was again listed in Slater's trade directory for 1844.

John Junior died before he had a chance to get into the 1841 census, but from the age given on his burial entry we know that he was born in about 1795 and was thus some 25 years younger than John Senior. In 1822 he advertised in Pigot's Directory as "John Williams jnr." and in 1824 he and his wife Anne had a daughter, whom they baptised Anne, the first of five recorded children. Anne (his wife, that is) died in 1836 shortly after the birth of their last child, Alfred, and John died three years later in 1839, but not before he had made a certain Emma Clarke pregnant. Their son was born on April 18th, the day before John Williams' burial. The parish register baptism entry for June 28th reads:

> "Born April 18th Benjamin Cross, Posthumous Son of John Williams deceased and Emma Clarke High St Watchmaker & Late Parish Clerk."

This is not the last we hear of John Junior. On August 22nd the parish register records the burial of Edward Williams (whose birth has gone unrecorded), "Son of John W^{ms} Watchmaker & Clerk, Deceased"; and in 1840, both John Williamses appear in Robson's trade directory.

A reference to one of the John Williamses is to be found at second hand in an article entitled "Early Victorian Denbigh" by E.P.Williams[29]. His article is based on "Denbigh as I remember it", the reminiscences of Abel Anwyl written down in 1909, which describes amongst other things the tradesmen working in the town. The reference which interests us is:

> "John Williams, known as Long John, was sexton of St. Hilary's and a watchmaker on the site of the present GPO."

The present GPO to which Abel Anwyl refers is still the Post Office. John Senior's address in 1840 is given as Hall Square and John Junior's in 1839 was Market Place. Of the two, then, it is more likely that Abel Anwyl is describing John Junior. It is a pity that it is impossible to know now whether the epithet of Long John was intended literally or ironically.

---

[28] In fact, this is not entirely accurate. Dr. Peate's book "Clock and Watch Makers in Wales" records the third Thomas Bartley as working in Pwllheli and notes three long-case clocks with his name on the dials. However, not one of his Denbigh clocks has surfaced as yet.
[29] Denbigh Historical Society; vol 22; 1973

Between them they worked in Denbigh for 57 years and two clocks have recently come to light signed J. Williams. The signatures are slightly different, since one is a straightforward J. Williams, whereas the other is Jnº Williams. We should not read too much into this, however, and both clocks may well be from the same shop. The older of the two clocks is a long-case clock (figure 25) in a pleasant and nicely decorated oak case. The well-painted dial is square and in very good original condition and its style suggests a date of about 1820. This is confirmed by the false-plate (a cast iron plate between the dial and the movement which is standard on white-painted dials) which bears the imprint of Walker & Hughes, a firm of dial makers who were in business in Birmingham from 1811 to 1835[30]. This very pretty clock could therefore have been sold by either man.

[Figure 25]

I have seen two other long-case clocks, both with square dials. One of these clocks (figures 26 & 27) is certainly by John Williams Junior, because the dial is signed with this name and the other (figures 28 & 29) is almost certainly by the same maker because, although it is signed only John Williams, it is housed in an almost identical case. Indeed the only difference which I can detect between the two cases is the shape of the top of the trunk door. Particularly striking are the square reeded columns to hood and trunk with the same horizontal box and ebonised fruitwood stringing. The movements are identical. However, the hands are of a different pattern, as are the dials, one having a false-plate with the name Wilkes & Son, whereas the false-plate on the

---

[30] White Dial Clocks; Brian Loomes; 1981; p. 56

[Figure 27 - close-up of a finial]

[Figure 26]

second clock has the name T.Hobson & Sons. Although the exact pattern is different, the style of the two dials is very similar, dating from around 1825 - 1830. The two clocks are a fine example of how the clockmaker matched up movement, dial, hands and case for his customers who could have chosen from what the clockmaker had in stock or ordered a pattern, perhaps from a catalogue or from a dial he had already seen in, say, a neighbour's house.

[Figure 29 - close-up of a finial]

[Figure 28]

Yet another clock (figure 30) is more unusual. The dial is signed J. Williams, Denbigh and dated 1838. It is a wall clock and has what is known as passing strike; in other words it strikes just once on the hour, whatever the hour. It is driven by a large weight and the case, although not as tall as a long-case clock, is long enough to allow the clock to go for one week between windings. The case is just under four feet high, about 14 inches wide and 7 inches deep. The single winding hole is placed to the right of the dial centre rather than directly beneath. The movement is similar in size and construction to a normal long-case movement but with none of the striking work which would be required for a normal long-case clock; indeed, to all intents and purposes it is half of a long-case movement. Thus all the signs are that John Williams designed this clock specifically for his customer and so it represents a notable exception to the custom of assembling a clock from standard components. This is corroborated by the fact that the clock was until recently in a local chapel. The chapel records have unfortunately not survived so we do not know if the clock was made for a special occasion. Its provenance may also help to explain the unusual passing strike, since a single blow every hour would have served as a discreet and tactful reminder to the preacher of the passing time and the waiting Sunday dinner.

[Figure 30]

A clock which was made by neither John Williams Senior nor John Williams Junior is the Denbigh Town Hall clock. At first sight this may seem to be a singularly useless piece of information. However, John Williams Senior did offer to make the replacement for Robert Farr's clock and the letter which he sent to the town council is to be found amongst the town clerk's papers:

"   Denbigh 1st Jan'y 1835

　　　To the Aldermen and Councillors of the Borough of Denbigh.

　Gentlemen / I am given to understand that the repairs and care of the
　Clock on the Town Hall is to be given under the Care of a Clock
　Maker. In that case I humbly solicit the favor of the Appointment and It
　shall be my endeavour to give you and the Town in General
　Satisfaction Should I succeed in getting the favor for which I shall
　always feel Grateful.
　　　　　　　　　　　　　　　　I am Your Most Obedient Servant

　　　　　　　　　　　　　　　　　　　　　　　John Williams Sen[r]
　　　　　　　　　　　　　　　　　　　　　　　Clock Maker"

In the event, he was not entrusted with the making of the new clock and, although all further correspondence on the matter has disappeared, we know that the clock which now stands above the library was made by the firm of Joyce of Whitchurch since the name Thomas Joyce & Sons and the date of 1847 are cast into the clock frame. The proximity of the well-established business of the Ruthin branch of the Joyce family may have helped in securing the deal for the Whitchurch firm.

The third John Williams is fortunately easily distinguished from his two predecessors, because he was known as John G. Williams (his middle name was Gabriel). He was born in Beaumaris in about 1821, married Elizabeth and appeared in Denbigh in 1841, living in what was then called Park Lane, now known as Park Street. His trading address in 1850 was High Street and in 1856 Hall Square. After 1856 he disappears from the scene.

Two of his watches are in the Welsh Folk Museum in St. Fagan's. The first is in a handsome but sadly damaged tortoise-shell outer case enclosing a silver inner case which is hall-marked Birmingham 1841. The standard verge movement with fusee is very similar to those sold by Richard Griffith (see chapter V) and indeed the case was made by the same firm (JH). The watch paper inside the case proclaims: "A General Assortment of Gold & Silver Watches Always on Hand. Gold Rings"; the inside of the case is scratched with the initials EW and dated AD 1842 and a small paper disc records the information: "Edward Williams, Castle Street, Rhuddlan 2". Presumably these are the name and initials of the watch's first owner. A second watch of similar type but this time with a silver outer case and by a different casemaker (BK) has a partly obliterated hall-mark which is almost certainly 1859[31].

Contemporaneously with John G. Williams, Joseph Duke worked in Denbigh for a few years. Our first mention of him is in Slater's 1844 trade directory and in the six years 1844 to 1850 he and his wife, Elizabeth, had five children, including twin boys. According to the 1851 census, both Joseph and Elizabeth were from Liverpool, they were both 37 years old, and the family lived in Vale Street. In the same census another Joseph Duke, also married to Elizabeth but both aged 62, were living in Chapel Street.

[31] Dr. Peate places this watch in the 1840s; op. cit., p. 83

These Dukes were from Chester and Joseph was a goldsmith. Presumably the two Josephs were father and son. Joseph Duke (the watchmaker) is not mentioned in the next trade directory (Slater's 1856), so it would appear that he and his family's stay in Denbigh was of relatively short duration.

Whereas the Dukes had little impact on the town, David Price Jones definitely made his mark on Denbigh. He came from Newcastle Emlyn and lived in 26 Park Street for some time before moving to 4 Hall Square (in the 1891 census he is listed as living in 8 Hall Square, but this conflicts with all other references, both before and after).

The reason for David Price Jones coming to Denbigh was that his wife, Mary Alice, was born there, but their first child, Fanny Mariah, was born in "Aberdar (Glam)" according to the 1871 census. Presumably Mary Alice had been married before, since the census records a stepson, Robert George Kelly, whom we have already discussed in this chapter. In 1891, three generations were living in the family home: David Price Jones and his wife, Robert Kelly, and Robert Kelly's son, David.

The "Dyn o'r Stryd" (Man in the Street), who described David Lloyd Griffith, also had his memories of David Price Jones:

> "A very small man, one of the characters of the town and one of the most gentlemanly of men you could possibly meet. He kept a good stock of clocks and watches and his window was dressed attractively. Inside the shop you could imagine on any day that that was the day the business opened. I have already said he was gentlemanly and quiet, but like many of us he had a great weakness. Sometimes mock auctioneers would come on fair days to sell watches and other things made specially for this kind of business. In Denbigh they sold many of their goods through their ability to overpraise poor things which were nice to look at.
> Once, a relative of mine left one of these watches in our house one Wednesday evening. My mother commanded me to take it to Mr Jones, Hall Square, to be repaired next morning before school. I put it in Mr Jones' hand and told him the name and address of its owner. He looked at it for a second or two and I noticed a frown creeping across his face. "Goodness gracious." Before I realised what was happening he had come rushing from behind the counter towards me. He planted the watch in my hand and a storm of abuse broke from his lips and I from surprise was unable to move. He put his hand on my shoulder and said: "Get out, boy, for your life, I will never mend a cheapjack's watch." He opened the door and pushed me right outside, so hastening its closure. I told my story to my father at lunchtime and he laughed loudly with much enjoyment. Despite my surprise, I experienced a mixture of feelings about the watch in the same day. This is the kind of man Jones the Watchmaker was. Such a wild temper, but what was remarkable about him was that he could quickly master it."

VII

From the middle of the 19th century a new family came to play a leading role in the clockmaking scene in Denbigh. This was the Keepfer family and they immigrated from the Black Forest. The exact date of their arrival is uncertain. According to Dr. Peate[32], the Keepfers arrived in 1850, and the firm's 1944 letter-heads agree with this; however, the letter-heads of 1908 claim that the business was established in 1852 and the obituary printed in the *Denbigh Free Press* in July 1912 states: "Wilhelm Keepfer and his father first came among us in 1852, he at the age of 17". This is not the only uncertainty regarding the family's arrival in Denbigh; as we shall see, they appear to have spent a short time in Abergele when they first arrived in North Wales.

The Keepfers (their German name was Küpfer) were just one of many families to flee the German states after the 1848 revolution and it was inevitable, since clockmaking was so wide-spread in Southern Germany, that many clockmakers would bring their skills to their new homes. Indeed, Dr. Peate[33] makes it clear that the Keepfers were part of an important influx of foreign expertise into the British - and more particularly the Welsh - trade.

The Keepfers were to become a highly respected family in Denbigh. In fact, our first direct reference to the family is in a newspaper article in the *Denbigh Journal* in 1854 which lists contributors to the Denbigh Soup Charity: "Keepfer, Mr. ... watchmaker ... 1/-". It is worth remembering that the Keepfers who donated their shilling to the charity were themselves refugees who had left their home town in Germany only a couple of years earlier. The Keepfer referred to here is Johann Keepfer who anglicised his name to John and is mentioned in Slater's 1856 trade directory as "Keepfer, John & Son ... High Street", the son being the by then 21-year-old Wilhelm. (Throughout, I shall refer to the first William as Wilhelm, and the second as William - after all, the first William officially stuck to the German form of his name up to, and including, the 1881 census; and it will be less confusing).

There is no record of Johann's death and burial, and it may well be that he and his wife Teresia [sic] were unable to settle in Denbigh, returning to Germany once they had seen their son established in Wales. In contrast, there is no lack of testimony to Wilhelm's achievements. Perhaps our best starting point for an assessment of his life would be the letter published in the *Free Press* as a tribute to his memory in the issue which records Wilhelm's funeral and which I will quote in its entirety:

> "Sir, - In your reminiscence in your last week's *Free Press* of my very old friend, Mr Wilhelm Keepfer, you state that he commenced business in High-street. This is not correct. Mr Keepfer and his father, and a brother, if I remember well, commenced in a small shop at the left hand side in Portland-place.
> The lock-up and the stocks were close together, "Hukyn Twrch," who used to be placed in them for making too free use of "John Barleycorn."
> Yes; he was instrumental in starting the Roman Catholic cause. This was in a large room over a Cooperage Shop in the Factory Buildings. He kept true to his Church, and was a very bright example in toleration

---
[32] Op. cit., p. 57
[33] Op. cit., p. 25

towards those who differed from him in religion and politics.
He was also a honest and conscientious tradesman. Fifty years ago I
purchased an eight-day clock from him, and better value for the money
never was purchased.

THOMAS PRYCE.

Clwyd House,
Westminster Park,
Chester."

This letter admirably sums up the triple achievement of Wilhelm's life: in his trade, in politics and in his championing of the Roman Catholic church.

The registers of the Society of Jesus at Pant Asaph chronicle his family life. He seems to have met his first wife, Elizabeth née Booz, in Denbigh. She too was a refugee; her place of birth in the 1891 census is given as Hassen. She was his senior by four years. They married in 1864 and had four children in all. The record of the birth of their first child is typical of the entries:

> "Die 24 Augusti 1865 nata et die 27 Augusti 1865 baptizata est Maria Teresia Küpfer filia Gulielmi Küpfer et Elizabeth Küpfer (olim Booz) conjugum a me Joanno O'Reilly S.J.St. Ap."

William[34] was born and baptised two years later in 1867, to be followed by two more daughters, Elizabeth and Winifred.

By 1853, there were already plans to build a Catholic church in Denbigh, and no doubt the Keepfers were involved in this project. The *Denbigh Journal* for October 11th, 1853, in an article entitled "Popery in Wales", deplores the spread of Catholicism in Wales and states: "Last week they bought a piece of ground in the town of Denbigh, on which a chapel is to be built shortly."

Throughout his life, Wilhelm campaigned vigorously for equal rights for Catholics. The following report from the *Free Press* of October 29th 1887 on what was obviously a lively council meeting gives us some flavour of the debates in which he became involved:

> "A protest against Intolerance.
> Mr Keepfer supposed that the members were aware that he had attended every meeting of the Council on these byelaws, and he had fought several hard fights as to some of the points. He urged that they should strike out the hours which the man [the grave-digger] was to be employed from 9 to 6, for he should be very sorry to put the Town to so fearful an expense, as would be the case if they had to pay the salary mentioned, and it behoved them to be careful in spending the ratepayers' money. He now made a claim. They had nearly completed the byelaws, but he now had a claim, as he had done before, that was for somewhere for the Catholics to be buried (hear, hear). He acted on

---

[34] The name on his birth certificate is Wilhelm Leopold.

the principle that they should do to others as they would be done by. Mr Gee had had his go and it was his turn now. Was this corporation going to deviate from the course followed by every other, such as Wrexham and Mold (hear, hear). He appealed to Mr Gee, is Denbigh more intolerant than any other place, or what was the matter that Denbigh was different to everywhere else? Were they as Catholics to be treated as outcasts? The claim of Mr Gee for the Nonconformists had been met in full, and was not the claim of the Catholics to be considered? Perhaps Mr Gee would help a plan, by which his petition may be granted, so that his religious body could have a portion of the cemetery. They were the poorest of the lot, and if they wanted to act towards the weakest sect liberally let them accede to his claim.

The Mayor told him that it was open to the new Council to accede to such application, and he should certainly make it. He took his word that he could open the question at a future time.

Mr Gee said he could not hear, nor quite understand Mr Keepfer's remarks, and so could not reply. It was out of no disrespect.

Mr Keepfer (loudly): Well, Mr Gee, I give notice that I intend moving that a portion of the new cemetery be set aside for Catholics. Do you understand that Mr Gee?

Mr Gee: I do.

Mr Keepfer: Very well, I hope you will support it then.

The Mayor did not see that the question now raised touched in the least degree upon the issue. It had been resolved at a previous meeting that no portion of the new cemetery be consecrated. Mr Keepfer would have to apply to the Council to rescind that resolution, and move that some portion of the cemetery be consecrated for Catholics. None of the byelaws would however prevent Mr Keepfer moving such a motion. The Catholics, however, were no worse off than Churchmen for whom no separate portion had been set aside..."

A memo in the register in Pant Asaph shows that Wilhelm's efforts were not in vain:

"Memo Oct. 6th 1889
The portion of the new cemetery which was secured to us through the efforts of Mr. Wm. Keepfer was solemnly blessed according to the form prescribed in the Roman Ritual on Sunday October 6th 1889, in virtue of faculties received by me from the Bishop for this ceremony."

It should not be thought, however, that Wilhelm used his position as Councillor only to further the Catholic cause.

He was elected Councillor in 1886, presumably the first Catholic to be a Council member, and remained in the Council until 1895, when he was defeated at the poll. In 1891-1892 he was Chairman of the Highway Committee and in 1894 he failed by a single vote to be elected Mayor. However, three years after losing his seat on the Council, he was elected Alderman and served until 1901 when he was again defeated.

The attendance records show that he was a most conscientious member of the Council, and the *Free Press* reports of Council proceedings suggest that he was always ready to speak his mind, and that Council meetings were usually entertaining when he was there:

> "Mr David Jones said that the complainant should have approached the officials and not gone behind their backs to the Council members.
> Mr Keepfer: I am much obliged to you, Mr Jones - much obliged to you. But the man never told me about it, I am, however, much obliged to you all the same. The people of Henllan told me about it and as I am the 'Champion of Henllan' as they call me, I take it up (loud laughter and cries of 'oh, oh').
> Mr Lloyd: The what, did you say?
> Mr Keepfer: The champion of Henllan (renewed laughter).
> Mr Lloyd: Well that is good certainly (laughter).

Altogether, Wilhelm was involved in many aspects of Denbigh life: he was a member of the School Board for many years, he helped to found a Chamber of Commerce and Agriculture, he was on the committees of the Free Reading Room, the Denbigh Floral, Dog and Poultry Show, the Winter Show, the Society for the Prosecution of Felons and the Management of Denbigh Castle.

One might be forgiven for wondering how Wilhelm found the time to run his business which is, after all, the main subject of our interest. He was undoubtedly a shrewd businessman and established his shop at 46 High Street as a reputable concern from the outset[35]. It provided work for himself and his son and a series of apprentices, all from Baden in Germany. In 1871 Phillip Meisinger (age 16) was listed as working for him; in 1881 Frederick Gauter (age 16) and Jacob Bauer (age 24) were both living above the shop; and in 1891 Jacob Bauer had changed his name to Bower and become bi-lingual, which in the context of the census means English and Welsh rather than English and German! In fact, Jacob Bauer must have come to Denbigh in the mid 1870s, as he was already a "Watch & Clock Maker" by the time of the 1881 census. Jacob claimed the right to vote in 1885 by virtue of the room he occupied over the shop; however, his claim was rejected.

Jacob Bauer followed the example of his employers in anglicising his name to Bower and settling in Denbigh for the rest of his life. A row of houses in Lenton Pool bears his name and his obituary in the *Denbigh Free Press* on August 9th, 1924, shows his qualities:

> "The death occurred on Tuesday, of Mr. Jacob Bower, of "Cynhasedd", Lenton Pool, Denbigh, at the age of 68 years. Deceased, who was of German extraction, came to this country when quite a young man, and for the last 19 years had been in the employ of Mr. W.Keepfer, watch-maker, High Street. He was a highly skilled and much valued servant, being a master of his particular branch of work. He had been in failing health for the last twelve months, but followed

---

[35] Our "Dyn o'r Stryd" said that he "had an extensive trade because he was a man in whose word you could trust at all times".

his employment whenever possible. While in Germany he was a
member of the Lutheran Church, and at Denbigh for many years
attended the English Church & Sunday School. Deceased was of quiet
disposition, and was held in high esteem."

The following week, two letters appeared in the *Free Press* paying further tribute. The first speaks with appreciation of his success in learning Welsh:

"The deceased gentleman, although not of Welsh nationality, could
speak the language fluently, his knowledge being wide and deep."

In the second letter, written by a Mr. J.H.Howard of Worksop, we read again of Jacob Bower's quiet personality when he is described as a "student in his own quiet, unassuming way." Equally interesting is the reference to the difficulties he had obviously experienced during the Great War. The letter writer mentions "calumny and abuse" and points out the irony of Jacob Bower's situation:

"He left his native country to avoid conscription - few men held
militarism in greater detestation."

When we read of Jacob Bauer (to revert to his original name) in these terms we see his great achievement in copying the example of his employer and integrating himself into the fabric of local society even though his attempts were evidently not totally successful when the tensions of the Great War made themselves felt.

The Keepfers sold thousands of clocks over the years, and a typical example of their wares is shown in figure 31. It is a sign of the times, and of the approaching crisis for British clock manufacturing, that the clock in question is an American import. It was the belief of the British makers that the public would pay high prices for high quality clocks, hand finished and in well-made cases. The philosophy in almost all other countries was that cheap and cheerful clocks would represent good value for money and sell well. This led to the development of mass-produced clocks in France, Germany and the United States. The American clocks were particularly good value for money. The clock we are discussing, housed in a mahogany case with some nice inlay, is described on the maker's label as a "Superior 8 day clock. With Extra Bushed Movement. Warranted if Well Used for home & abroad". The British refusal to produce what they considered to be inferior clocks looks a little short-sighted when one considers that this clock is still performing quite happily as it approaches its 100th birthday. It has none of the refinement of a British made clock of the same period, nor is it as well-finished as the mass-produced French clocks of the 19th century; nevertheless, it could well qualify for the approval expressed by Thomas Pryce in his letter to the *Free Press* after Wilhelm's death.

To quantify the relative prices which Victorians paid for their clocks we can consider the wholesale value of the clocks recorded in 1868 in the storerooms of the well-known firm of Camerer Cuss & Co[36] in London. In their books, the average value (converted into decimal currency) of the various categories of clocks which they stocked was as follows:

---

[36] I am indebted for this information to Mr. Terence Camerer Cuss who details these costs in his book "Camerer Cuss & Co - The Bicentenary".

[Figure 31]

[Figure 32]

[Figure 33 - detail of arch scene]

English dial clock	£6.09
French carriage clock	£5.80
French marble clock	£3.35
Cuckoo clock	£1.57
German mantel clock	.91p
American clock	.53p

Retail prices would have been about 60% higher than this. With such a huge price differential it is hardly surprising that imported clocks were popular.

The Keepfers also sold traditional British clocks and there is a Keepfer long-case clock in the Welsh Folk Museum and another in private ownership in Denbigh; both are typical of the final years of the Welsh long-case clock, with very solid, wide cases, painted dials and factory-made movements. The local clock (figures 32 & 33) has a lively shooting scene in the dial arch, and the quality of the painting is rather better than many of the last dials to be produced.

Three of the clocks I have seen are signed W. Keepfer & Son. It is not clear exactly what date this represents, since the trade directories never use the & Son. It would be reasonable, though, to assume that William joined his father as a fully-fledged member of the business in about 1887 when he was 20 and that the & Son appeared on the dials at about this date. All of these three clocks are wall clocks, two of the Anglo-American variety[37] and the third a round schoolroom type clock.

What may seem surprising, in view of the Keepfers' continuing contacts with their native Germany, is the fact that they do not appear to have specialised in Black Forest clocks. Certainly none of the publicity material that I have seen suggests that the firm went out of its way to publicise German clocks and if they had sold a sizable number of Black Forest clocks, presumably some would still survive.

As we saw earlier when discussing the transition from the 18th to the 19th century, the work of a clock or watch maker did not involve any real clock or watch making! Even though W. Keepfer & Son described themselves as "Manufacturing Watch and Clock Makers", their work would have involved repairs, servicing and sales. An invoice dated 1908 (figure 34) reminds us of another aspect of their work, namely winding clocks on contract. Mr. A. Ffoulkes-Roberts paid the Keepfers 10/6d for winding his office clocks for a year. The contract ensured that the clocks were kept accurate at weekly intervals, and that any problems would be noticed well before the affected clock ground to a halt[38]. Then, as now, owners tended to have their clocks serviced when they stopped rather than when they actually needed servicing, by which time considerable damage could already have been caused.

Examination of the bill shown in figure 34 is quite rewarding for what it illustrates about the typical goods sold by the Keepfers. Two watches, a small alarm clock and a black "marble" clock manufactured in France or the USA stand beside a variety of cut glass and silver goods. The Keepfers themselves were at pains to emphasise that they

---

[37] Anglo-American designates a clock with a British-style case (often manufactured in the United Kingdom) which houses an American-made movement.

[38] This service continued to be available in London until the end of the 1970s.

were jewellers and silversmiths as well as clockmakers. Moreover, they sold "All kinds of spectacles" and in Porter's 1886 trade directory Wilhelm Keepfer described himself as "Watchmaker, jeweller and optician".

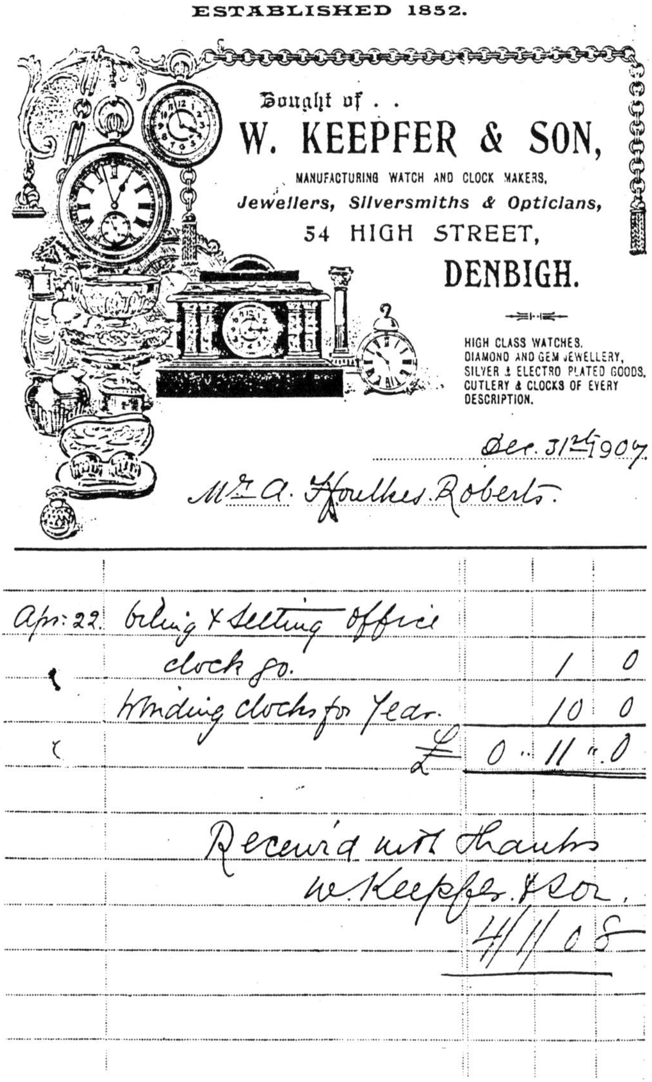

ESTABLISHED 1852.

Bought of . .

W. KEEPFER & SON,

MANUFACTURING WATCH AND CLOCK MAKERS,

Jewellers, Silversmiths & Opticians,

54 HIGH STREET,

DENBIGH.

HIGH CLASS WATCHES.
DIAMOND AND GEM JEWELLERY,
SILVER & ELECTRO PLATED GOODS.
CUTLERY & CLOCKS OF EVERY DESCRIPTION.

Dec. 31st 1907.

Mr. A. Foulkes Roberts.

Apr: 22.	Oiling & Setting Office clock go.	1	0
	Winding clocks for Year.	10	0
	£ 0 .. 11 .. 0		

Receiv'd with thanks
W. Keepfer & Son.
4/1/08

[Figure 34]

Since the scope of this book does not extend into the 20th century we shall not follow the Keepfer family beyond 1900. Instead we shall reflect upon Wilhelm's clever juggling act in which he retained close ties with his German roots while becoming totally integrated into the Welsh community he had joined. He called his home in Broomhill Lane "Schwarzwalder Cottage" and for Queen Victoria's Jubilee in 1887, having decorated his shop with red, white and blue and many Union Jacks, added a banner with the inscription "Lang Labe[39] Unsere Gnädige Königin" (Long Live Our Gracious Queen). His apprentices, as we have seen, came from Germany. On the other hand, he also took up an entirely new and thoroughly un-German activity: his name features in the sports pages of the *Free Press* during the summer months as a member of the Denbigh Cricket Team. In 1887, for example, he took two wickets against Abergele but scored 0. He managed 5 runs a fortnight later against Mold but was out for a duck in the return match in August.

In conclusion, the newspaper tribute to him after his death sums up the respect in which he was held:

> "His public actions were always characterised by straightforwardness,
> and his integrity in both public and private life, was beyond suspicion".

The fiercest competition to the Keepfers must have come from the Joyces whose family history and impact on the Vale of Clwyd we saw in an earlier chapter. By the time the Keepfers had established their business, Henry Joyce was also working in Denbigh and doubtless there was a healthy rivalry between the two firms.

A description in the *Free Press* of the Denbigh Christmas Show and Market in 1882 singles out Henry Joyce's display for particular praise:

> "Mr Henry Joyce made an effective display of much jewelry of all kinds, clocks of special make and of use for the kitchen or drawing room. A great attraction here was a mechanical clock under a glass shade which, when wound up, caused a musical box to play a variety of tunes, troops of soldiers to march from Egypt, a windmill to turn, a ship to sail, etc.
> Mr W. Keepfer had an excellent show of jewelry and articles for Christmas, also a musical clock and other attractions. Excellent stocks of jewelry, clocks, etc. were shown by Mr Griffiths and Mr D.P.Jones."

Both firms had a keen sense of salesmanship and Henry Joyce used a picturesque engraving of Denbigh Castle to decorate his watch case papers (see figure 12). It is interesting to note that the Keepfers (exactly like Griffith & Son - figure 24) chose to ennoble their watch-case paper (figure 35), this time with a royal crest and its twin mottos, "Honni soit qui mal y pense" and "Dieu et mon droit". Their publicity also took the form of a public clock which was placed in the wall above the shop. Figure 36 shows Johann Keepfer's grandson, William, standing in the shop doorway and the clock face can be seen above him, although the time is not legible in the photograph. A second dial faced into the room above the shop for the benefit of the inhabitants.

---

[39] Correctly, this should read "Lang Lebe ..."

[Figure 35]

[Figure 36 - Picture loaned by Mrs. P. Evans]

VIII

The clockmakers of Denbigh have so far dominated this book by virtue of the fact that the town of Denbigh played a leading role in the Vale during the period under consideration. However, during the first half of the 19th century a busy new town suddenly sprang up. The town was, of course, Rhyl and its development was the result of the arrival of the railway.

The railway was opened in May 1848 and the population grew from fewer than 300 in 1801 to ten times that number in 1861. New though the town may have been, it nevertheless needed all the services which an established town could offer and obviously high on the list of requirements was a clockmaking trade with its ancillary branches of jewellery, barometers, cut glass and the like. The clockmakers came from near and from relatively far to provide these services and an analysis of the origins of the Rhyl clockmakers splits about half and half between local men and men from further afield.

At one extreme is a maker such as Charles Matthews, who was born in St. Asaph. In 1881 he was living with his father on the family farm, aged 22, unmarried and a "Watchmaker". They were a local family. Frederick Matthews, the father, had also been born in St. Asaph and his wife, Elisabeth, was from Ruthin. Frederick described himself as a "Farmer of 4 acres & Watchmaker" and one supposes that he must have fitted in his watchmaking skills, perhaps self-taught, around the demands of his smallholding. This of course was how the extremely successful Black Forest farmers had built up an extensive parallel business as winter-time clockmakers which in turn had led to mass produced manufacturing in the 19th century. Charles obviously left the family farm a few years later to re-appear, still unmarried, as a lodger at 24 Water Street in Rhyl where he now described himself as a "Jeweller".

The other extreme is represented by Arthur Merridew who appears in the 1871 and 1881 Rhyl censuses. He was born in Coventry, as was his wife, Emma, and their first daughter, Emma E., who was 14 in 1871. At some time between 1857 and 1862 the family had moved to Stretton-upon-Dunsmore, in Warwickshire where their first son, Frederick, was born. However, they cannot have stayed there very long because their second son, Charles, was born in Dinstall, in Staffordshire, in 1866 or 1867. Arthur Merridew described himself as a "Manufacturing Goldsmith and Watchmaker" and was a successful businessman with interests outside the sphere of horology. In "Rhyl and Round About"[40] J.W.Jones had this to say about him: "A.W.Merridew was a jeweller in Bodfor Street who built many of Rhyl's business premises. When he retired he went to live in Colwyn Bay."

It is typical of this era that Mr. Jones describes Arthur Merridew as a jeweller. By the last quarter of the 19th century the public perception of the clockmaker had changed with the change in the clockmaker's role so that his involvement with clocks and watches was no longer of primary importance. This was probably inevitable as clockmakers had evolved from making clocks, to assembling them and finally to selling ready-made clocks in which they had had no hand at all, beyond ensuring that the dial advertised their names. The fact that Arthur Merridew's two older children, Emma and Frederick, are described respectively as "Assistant" and "Watchmaker" shows how the

---

[40] "Rhyl and Round About", J.W.Jones, Clwyd Press, 1976

business was run, with the father masterminding the operation while the son carried out the clock and watch repairs and the daughter served in the shop.

Just for the record, the origins of the Rhyl clockmakers who were recorded in the censuses from 1851 to 1891 were, in alphabetical order, Bangor (Gwynedd), Birmingham, Caernarvon (3), Colchester, Coventry, Llangynog (Powys), Llanrwst, London, Rhuddlan, Rhyl (4), Runcorn, St. Asaph (2), Salford and Whitby.

In Arthur Merridew we have seen a successful businessman, but not all clockmakers were so fortunate. A case in point would be Owen Roberts whom we glimpse three times in the last twenty years of the century. In 1881 he is 24 years old and married with a one year old baby son, Enoch. According to the census he was born in Caernarfon, his wife was born in Rhuddlan and the baby was born in Denbigh. Ten years later he is working in Rhuddlan High Street, where his entire household, consisting of 9 people, lives in what is described as a 2 roomed house and his fourteen year old stepson, William Middleton, has appeared from somewhere. Owen is said in the census to be Welsh speaking rather than bilingual and he and his family will move from Rhuddlan to St. Asaph in the next few years since an entry in the parish register there records the baptism of Blodwen in 1898. It would be hard to imagine from what we know that Owen Roberts lived anything other than a very difficult existence, and this is supported by the fact that he seems to have worked in four different towns in the space of twenty years. This contrasts with what we observe as the successful families in clockmaking who, once they were settled in a town, stayed there for a good number of years, frequently passing the business on to their children.

Rhyl is, of course, known as a seaside resort, and the censuses of the second half of the century have caught a couple of holiday makers who, although they cannot be classed as Vale of Clwyd clockmakers, may well fit into someone else's research. William Henry was visiting his colleague William Hughes at 7a High Street in Rhyl in 1891. Both men were bilingual but there was twenty years between them. William Henry had been born in Wrexham whereas William Hughes was a St. Asaph man. In nearby Rhuddlan, a William Halliday was staying in the White Horse Inn in 1871. He was aged 59 and had been born in London.

Neither of these two men is recorded in any of the reference works which I have consulted and I mention them primarily to ensure that their names, which may be of interest to someone somewhere, do not go unrecorded.

When we were discussing the Griffith family of Denbigh we saw that they had sidelines to supplement their income. The same was equally true at the end of the 19th century when clockmaking had come to mean something far less horological. A couple of examples will suffice to show that the clockmakers of Rhyl were happy to turn their hands to whatever might prove rewarding. Arthur Merridew appears to have been involved in property development and this brought him a presumably happy retirement in a desirable residence in Colwyn Bay. The less entrepreneurial Richard Tomkies, who had also come to Rhyl from Coventry, described himself in the 1881 census as "Watchmaker & Tobacconist". As we have said, this was not uncommon, and a further example, this time from outside Rhyl, is to be found in St. Asaph where E.M Griffith is listed in the 1874 directory as watchmaker but also as proprietor of what is today Turner's Tearooms, opposite the cathedral.

IX

The first clock in this book was a public one and our work would not be complete without some discussion of a few public clocks in the Vale, ranging from the small to the magnificent. In the following pages I show illustrations of some of the public clocks which have contributed in some way to the life of the community in which they are placed. The role of these clocks to the community they serve is exemplified by the story attached to the clock of St. Asaph cathedral which is shown in figure 37. Apparently, the life of the city was perturbed not a little when the clock began to strike late as a consequence of the baby owls sitting on the minute hand and making their own adjustment to St. Asaph Mean Time. It is said that would-be passengers missed their buses and shoppers went without their purchases until the young birds had fled their nest and life could return to normal.

[Figure 37 - By kind permission of Mr. R. Thompson, F.I.B.P., F.I.P.S.]

[Figure 38 - Picture loaned by Mr. P. Smith]

Trefnant is a small village half-way between Denbigh and St. Asaph. This old postcard shows some inhabitants (and Jacko, the dog) outside what used to be the Post Office. The clock, supplied by J.P. Joyce of Denbigh, had been bought by public subscription and it has two dials, one inside and the other outside. The clock is still in the same building today.

[Figure 39 - By kind permission of the Clwyd Record Office]

The Clock Tower on the Promenade in Rhyl was built in 1948 and was presented to the town by Councillor and Mrs. R.L. Davies in December of that year.

[Figures 40 - By kind permission of the Clwyd Record Office]

The photograph of the Old Ruthin Town Hall (above) was taken after 1859, when the spire was added to St. Peter's Church, but before 1863, when the Town Hall was demolished. By this time, the church clock was in disrepair and so Ruthin was left with no public clock. The Joseph Peers Memorial Clock (below) was erected in 1883 on the same site to mark Mr. Peers' 50 years as Clerk of the Peace for Denbighshire. The shop of T.H. Rigby can be seen on the far left of the picture.

[Figure 41 - By kind permission of the Clwyd Record Office]

POST SCRIPT

Historically, this book ends at the end of the 19th century, although several of the makers whose work has been discussed continued to work into the 20th century. However, our exploration of the Vale of Clwyd would not be complete without at least some mention of the work of Mr. Thomas Roberts.

Tom Roberts was born on a farm in Llansannan in 1925. Through his mother's family he is related to David Griffith (Clwydfardd) and therefore to all the Griffith clockmakers stretching back to Richard Griffith and the eighteenth century. He attended Denbigh Grammar School, leaving at the age of 15 to become an apprentice fitter at a motor engineering works in Rhyl. This was during the war and he might well have travelled further away to do his apprenticeship had his parents not feared for his safety in the larger industrial towns and cities. On the other hand, some engineers were moving from the industrial centres and one of these, taking refuge from Derby where he had worked as a machinist for Rolls-Royce, came to Rhyl and worked alongside Tom Roberts who learnt much from his senior colleague. Just after the war, Tom Roberts did his National Service with the REMI, and two years later joined the Health Service. His work here involved engineering duties of all kinds, including instrument engineering. From 1969 onwards he was employed by what was then Flintshire County Council as a heating maintenance engineer. At the age of 53 he took early retirement and began to devote his time to clockmaking.

This was not the result of a sudden rush of blood to the head. Even at school he had been fascinated by clocks and had always been ready to volunteer to mend his friends' watches when they stopped. This had a triple advantage: it satisfied his curiosity about the insides of watches; it stimulated him to further experimentation; and some of the watches actually started to go again. Moreover, with no television to distract him, tinkering with the insides of mechanical devices was a thoroughly engrossing pastime. At this stage he was as interested in railways as in clocks and when he took the farm produce to Abergele market on a Monday morning he would regularly walk on to Pensarn where he could stand on the railway bridge and watch the mainline trains thundering past on the North Wales line.

As we have already said, he was interested in clocks from his schooldays onwards; however, it was not a foregone conclusion that on retirement he would turn all his attention to clockmaking and he was tempted to specialise in model engineering. Both activities offer the satisfaction of manual work and use similar techniques and throughout his working life he spent his spare time on both modelmaking and repairing and restoring clocks. He eventually chose clocks, feeling that model engines are ultimately no more than ornaments whereas clocks perform a useful function.

Since his retirement in 1978, Tom Roberts has repaired and restored thousands of clocks for owners throughout the Vale of Clwyd and further afield. In this respect he is little different from any of the other clockmakers who have worked in Denbigh over the centuries. It is, however, as a genuine maker of clocks that he has distinguished himself from the overwhelming majority of his predecessors.

The first clock which he constructed was the result of John Wilding's book "How to make an 8-day WEIGHT DRIVEN WALL CLOCK". Tom Roberts is only too happy to acknowledge the extent to which he is indebted to John Wilding for this and other workshop manuals in the series for the simple yet detailed explanations of the processes involved in making all the parts required. The successful construction of this first clock encouraged him to attempt further designs and so far he has completed eight clocks.

Tom Roberts has been awarded two Eisteddfod prizes for his clocks. He gained first prize for his 8 day Water Drum Clock, in which the speed of rotation of the movement is regulated by water draining between the vanes of a drum and second prize for his 8-day long-case clock. The case for the second of these two clocks was made by Andrew Lloyd, a local craftsman who also works in Denbigh. An original feature of the clock is the engraving of Denbigh Castle in the dial arch. Unlike other makers whose work we have discussed (the Minshulls, Christopher Sharrock and Richard Griffith), Tom Roberts did not spoil the work of professional dial engravers but was happy to leave well alone.

It is a fitting ending to this book to be able to record that clockmaking is still alive and well in the Vale of Clwyd. The last words will therefore be Tom Roberts's, describing the most satisfying moment in his career as "hearing the first tick of the first clock I ever made."

APPENDIX

Vale of Clwyd Clock and Watchmakers

All dates are working dates, unless indicated otherwise. Where the last reference is that of death or burial (indicated by the symbol †) the clockmaker is assumed to have worked up to his death.

Abbreviations:
AMD: Ancient & Modern Denbigh; John Williams
CWMW: Clock and Watch Makers in Wales; Dr. I.C. Peate
DFP: *Denbigh Free Press*
DJ: *Denbigh Journal*
HPR: Henllan Parish Register
RCA: Ruthin Churchwardens' Accounts
RhPR: Rhuddlan Parish Register
SJPA: Society of Jesus, Pant Asaph
Dir.: Trade Directory
DPR: Denbigh Parish Register
QSO: Quarter Sessions Order Book
RPR: Ruthin Parish Register
SAPR: St. Asaph Parish Register
SLCR: Swan Lane Chapel Register, Denbigh

* An asterisk indicates a maker mentioned in the text, with page reference(s); additional details are given for makers not discussed in the text.

ARGENT   William T.   [c.1816 - ?]   **RHYL**
born in Colchester - married to Frances
1861 (census: "Age 45; 10 West Parade; Watchmaker") - 1889 (Dir.: "19 High St")

* BARTLEY   Thomas (I)   [? - 1811]   **DENBIGH**
1798 (DPR) - † 1811 (DPR)   *pages 44-45*

* BARTLEY   Thomas (II)   [? - 1811]   **DENBIGH**
1809 (SLCR) - † 1811 (DPR)   *pages 44-45*

* BARTLEY   Thomas (III)   [c.1799 - ?]   **DENBIGH**
1851 (census) - 1871 (census)   *pages 44-45*

BATTY   Richard   **DENBIGH**
† 1814 (DPR: "lodger at the Red Lion")

* BAUER   Jacob / Jakob   [c.1857 - 1924]   **DENBIGH**
1881 (census) - † 1924 (DFP)   *pages 43, 54-55*

BODDINGTON William   [c.1828 - ?]   **RHYL**
born in Birmingham
1868 (Dir.: "24 Queen st") - 1891 (census: "Age 63; 28 Queen Street; Jeweller; single")

BROWN Alexander [c.1803 - ?] DENBIGH
born outside Denbighshire
1841 (census: "Age 38; Panton Hall Place; lodger; Watchmaker")

CLARK Augustus G. [c.1870 - ?] RHYL
born in Rhyl
1891 (census: "Age 21; 36 Kinmel Street; son; Watchmaker")

CORNEY John E. [c.1868 - ?] RHYL
born in Whitby
1891 (census: "Age 23; 46 Kinmel Street; boarder; Watchmaker")

* **COURTER** Edward [? - 1775] RUTHIN
1724 (RPR) - † 1775 (RPR) *pages 22-24, 28*

* **COURTER** William RUTHIN
1741 (RPR) - 1777 (RPR) *pages 22-24*

* **DAVIES** David DENBIGH
1738 (DPR) - 1740 (DPR) *page 29*

DAVIES David St. ASAPH
1840 (Dir.)

* **DAVIES** Edward [c.1863 - ?] RUTHIN
1881 (census) *pages 27, 43*

* **DAVIES** John (I) [c.1846 - ?] RHYL
1861 (census) *page 43*

DAVIES John (II) [c.1806 - ?] St. ASAPH
born outside Flintshire - married to Margaret
1841 (census: "Age 35; Lower Street; Watchm^r")

* **DUKE** Joseph [c.1814 - ?] DENBIGH
1844 (Dir.) - 1851 (census) *page 49*

EDWARDS Edward [c.1849 - ?] St. ASAPH
born in Flint
1891 (census: "Age 42; Temperance House, High Street; lodger; Watchmaker")

EDWARDS Gabriel [c.1821 - ?] RUTHIN
born in Ruthin - widower (1851); married to Catherine (from 1861)
1850 (Dir.: "Clwyd st") - 1881 (census: "Age 60; Borthyn 14, Llanfwrog; Watchmaker")

* **EDWARDS** Hugh [? - 1767] RUTHIN
1735 (RCA) - † 1767 (RPR) *page 29*

EDWARDS   William Harris   [c.1859 - ?]   RHYL
born in Carnarvon - married to Ellen
1881 (census: "Age 22; 8 Vaughan Street; Watchmaker")

ELLIS   David   [c.1767 - ?]   St. ASAPH
born outside Flintshire
1828 (Dir.) - 1844 (Dir.: "Denbigh st")

* EVANS   Richard   [c.1856 - ?]   DENBIGH
1871 (census)   *page 43*

* FARR   Robert   DENBIGH
1605 (DCM)   *pages 5-9, 19, 48*

* FOULKES   David   [? - 1736]   HENLLAN
1727 (AMD) - † 1736 (HPR)   *pages 30-31*

* GARNETT   Thomas   DENBIGH
1775 (DPR)   *page 29*

* GAUTER   Frederick   [c.1865 - ?]   DENBIGH
1881 (census)   *page 54*

* GREATRIX   Edward   DENBIGH
1844 (DPR)   *page 42*

* GRIFFITH   David "Clwydfardd"   [1800 - 1894]   DENBIGH
1868 (Dir.) - 1876 (Dir.)   *pages 34, 38-40*

* GRIFFITH   David Lloyd   [1849 - 1909]   DENBIGH
1871 (census) - † 1909 (DFP)   *pages 32, 38-40, 49, 59*

* GRIFFITH   E.M.   [?c.1845 - ?]   St. ASAPH
1874 (Dir.)   *pages 43, 62*

GRIFFITH   Hugh   DENBIGH
no date recorded (CWMW: "L.c. clock at Holyhead, Anglesey")

* GRIFFITH   Richard (I)   [c.1755 - 1835]   DENBIGH
1794 (DPR) - † 1835 (DPR)   *pages 28, 32-35, 39*

* GRIFFITH   Richard (II)   [1814 - 1841]   DENBIGH
† 1841 (DPR)   *pages 32, 37-39*

* GRIFFITH   Richard (III)   [c.1840 - ?]   DENBIGH, RHYL &
1861 (census) - 1881 (census)   *pages 32, 38-40*   **St. ASAPH**

GRIFFITHS   David   RHYL
1856: (Dir.: "Wellington rd")

* GRIFFITHS Edward   [c.1819 - ?]   RHYL
1861 (census)   *page 43*

HARRIS   Alfred   [c.1847 - ?]   RHYL
born in Russia - married to Annie
1881 (census: "Age 34; 6 High Street; Jeweller") - 1889 (Dir.: "6 & 7 High st")

* HARRIS   Mrs Annie Louise   RHYL
1887 (CWMW: "High Street [D]") *page 42*

HELSBY   Benjamin   [c.1835 - ?]   St. ASAPH
born in Rhuddlan - married to Sarah Annie
1861 (census: "Age 26; unmarried; Chester Street; Watchmaker (Finisher)") - 1864 (SAPR)

HEWLITT   Michael   St. ASAPH
married to Elizabeth
1870 (SAPR)

* HUGHES   William   [c.1835 - ?]   RHYL
1891 (census)   *page 62*

* JONES   David (I)   [c.1786 - ?]   RUTHIN
1822 (Dir.) - 1844 (Dir.)   *page 28*

JONES   David (II)   [c.1792 - ?]   St. ASAPH
born outside Flintshire - married to Margaret
1840 (Dir.) - 1841 (census: "Age 49; High Street; Watchmaker")

* JONES   David Price   [c.1827 - ?]   DENBIGH
1871 (census) - 1895 (Dir.)   *pages 43, 49-50, 59*

JONES   Edward   [c.1870 - ?]   RUTHIN
born in Treuddyn, Flintshire
1891 (census: "Age 21; 4 Well Street; Watch maker assistant" [Robert G. Joyce])

JONES   Edward D.   [c.1852 - ?]   RUTHIN
born in Wigan
1871 (census: "Age 19; 65 Well Street; Watchmaker's Apprentice" [Robert G. Joyce])

* JONES   Eleanor   [c.1831 - ?]   RHYL
1881 (census)   *page 42*

JONES   Griffith   RUTHIN
no date recorded (CWMW: "L.c.clock at Blaenau Festiniog, Mer.")

JONES   Henry (I)   DENBIGH
1895 (Dir.: "13 Vale street")

JONES   Henry (II)   [c.1838 - ?]   RHYL
born in London, St. Giles - married to Elizabeth
1881 (census: "Age 43; 31 Queen Street; Jeweller") - 1886 (Dir.: "31 Queen st")

JONES   John (I)   [c.1818 - ?]   RHYL
born in Rhyl
1861 (census: "Age 43; 72 Wellington Road; Watch Maker") - 1871 (census: "Age 53"; ditto)

JONES   John (II)   RUTHIN
no date recorded (CWMW: "L.c.clock, brass dial, at Ruthin, Denb."

JONES   John E.   [c.1851 - ?]   RHYL
born in Llangynog, Montgomeryshire - married to Mary
1891 (census: "Age 40; 14 St. Helen's Place; Clockmaker")

JONES   Richard   [c.1804 - 1830] RUTHIN
† 1830 (RPR)

JONES   Robert   [c.1770 - 1820] RUTHIN
1809 (CWMW) - † 1820 (RPR)

JONES   Thomas   DENBIGH
1794 (DPR) - 1816 (Dir.)

* JONES   William (I)   DENBIGH
1856 (Dir.)   *pages 43-44*

* JONES   William (II)   [c.1823 - ?]   RHYL
1851 (census) - 1874 (Dir.)   *pages 43-44*

* JONES   William (III)   RUTHIN
no date recorded (CWMW: "L.c.clock in the Llangefni district, Anglesey")
                                     pages 43-44

* JONES   William (IV)   [c.1817 - ?]   St. ASAPH
1850 (Dir.) - 1851 (census)   *pages 43-44*

* JONES   William (V)   St. ASAPH
1850 (Dir.)   *pages 43-44*

* JOYCE   Henry   [c.1837 - 1909] RUTHIN & DENBIGH
1851 (census) - † 1909 (DPR)   *pages 26-28, 40, 43, 59*

* JOYCE   Henry Edward   [c.1864 - ?]   DENBIGH & RUTHIN
1881 (census) - 1897 (RPR)   *pages 27-28, 43*

* JOYCE   James   [c.1801 - 1874] RUTHIN
1835 (Dir.) - † 1874 (RPR)   *page 26*

* JOYCE	John (I)	[1744 - 1809]	DENBIGH & RUTHIN
1774 (DPR) - † 1809 (RPR)		pages 24-26, 28	
* JOYCE	John (II)	[1784 - 1847]	RUTHIN
1822 (Dir.) - † 1847 (RPR)		page 26	
* JOYCE	John Parry	[1872 - 1948]	DENBIGH
not recorded until after 1900		page 27	
* JOYCE	Robert	[1788 - 1859]	RUTHIN
1828 (Dir.) - † 1859 (RPR)		page 26	
* JOYCE	Robert Griffith	[c.1829 - ?]	RUTHIN
1851 (census") - 1891 (census)		page 26	
* JOYCE	Walter Conw(a)y	[1840 - 1910]	RUTHIN
1861 (census") - † 1910 (RPR)		pages 26, 43	
* KEEPFER	Johann [John]		DENBIGH
1854 (DJ) - 1856 (Dir.)		pages 51, 59	
* KEEPFER	Wilhelm	[c.1836 - 1912]	DENBIGH
1865 (SJPA) - † 1912 (SJPA)		pages 39, 43, 51-55, 57-58	
* KEEPFER	William	[1867 - 1945]	DENBIGH
1891 (census) - † 1945 (SJPA)		pages 52, 54, 57, 59	
* KELLY	Robert George	[c.1867 - ?]	DENBIGH & St. ASAPH
1891 (census) - 1894 (SAPR)		pages 42-43, 49	
LEIGH	William		RUTHIN
married to Sarah			
1894 (RPR)			
LLOYD	Edward		RUTHIN
1856 (Dir.: "Well st") - 1868 (Dir.: "Well st")			
LLOYD	Hugh	[c.1873 - ?]	RHYL
born in Runcorn			
1891 (census: "Age 18; 14 Sussex Street; lodger; Watchmaker")			
* MATTHEWS	Charles	[c.1859 - ?]	St. ASAPH & RHYL
1881 (census) - 1891 (census)		page 61	
* MATTHEWS	Frederick	[c.1824 - ?]	St. ASAPH
1881 (census)		page 61	
* MEISINGER	Phillip	[c.1855 - ?]	DENBIGH
1871 (census)		pages 43,54	

* MERRIDEW   Arthur W.   [c.1829 - ?]   RHYL
1871 (census) - 1883 (Dir.)   *pages 61-63*

* MERRIDEW   Frederick   [c.1861 - ?]   RHYL
1881 (census)   *page 62*

* MINSHULL   Henry   [? - 1753]   DENBIGH
1705 (DPR) - † 1753 (DPR)   *pages 10-14, 17*

* MINSHULL   John (I)   [? - 1778]   DENBIGH
1722 (QSO) - † 1778 (DPR)   *pages 10, 15, 18-20, 31-32, 45*

* MINSHULL   John (II)   [1739 - 1789]   DENBIGH
1758 (DPR) - † 1789 (DPR)   *pages 15, 18-21, 31-32, 45*

* MINSHULL   John (III)   [c.1763 - 1845]   DENBIGH
1786 (DPR) - † 1845 (DPR)   *pages 15, 20-21, 45*

PALETHORPE Harry   RHYL
1886 (Dir.: "31/2 Bodfor st")

* PARRY   John   [c.1791 - ?]   RUTHIN
1828 (Dir.) - 1844 (Dir.)   *page 28*

* PERCH   James   [c.1811 - ?]   RUTHIN
1851 (census)   *page 42*

* PIERCE   Daniel David   [c.1845 - ?]   RUTHIN
1861 (census)   *pages 27, 43*

* READ   Snead   DENBIGH
1793 (DPR)   *page 29*

* RIGBY & RIGBY   DENBIGH
1895 (Dir.: "Bridge street")   *page 42*

* RIGBY   Elizabeth   [c.1852 - ?]   RUTHIN
1889 (Dir.) - 1891 (census)   *pages 42-43*

* RIGBY   Thomas Henry   [c.1865 - ?]   RUTHIN
1891 (census) - 1900 (RPR)   *page 43*

* RIGBY   William D.   RUTHIN
1886 (Dir.: "St. Peter's sq.") *page 43*

ROBERTS   David   [? - 1906]   LLANRHAEDR
1856 (WCMW) - † 1906 (DPR)

ROBERTS   Gabriel   RUTHIN
"Second half of the 19th century" (CWMW)

ROBERTS   Jesse   **[c.1842 - ?]**   **RUTHIN**
1886 (Dir.: "36 Clwyd st") - 1891 (census: "Age 49; 36 Clwyd Street; Watchmaker")

ROBERTS   John   **[c.1863 - ?]**   **DENBIGH**
born in Denbigh
1881 (census: "Age 18; 30 Beacon's Hill; Watchmaker (finisher)")

* ROBERTS   Owen   **[c.1857 - ?]**   **RHYL, RHUDDLAN**
1881 (census) - 1898 (SAPR)   *page 62*   **& St. ASAPH**

ROBERTS   William   **[c.1811 - ?]**   **RHYL**
born in Salford - married to Elizabeth
1874 (Dir.: "10 Market st") - 1881 (census: "Age 70; 15 Brighton Road; Clock Maker")

* SCANLAN   Selina   **[c.1855 - ?]**   **RUTHIN**
1871 (census)   *page 42*

* SHARROCK Christopher   **[c.1712 - 1783]**   **DENBIGH**
1759 (DPR) - † 1783 (DPR)   *pages 30-32*

* SOMNER   William   **RUTHIN**
1776 (RPR)   *page 29*

THOMAS   Edward E.   **[c.1868 - ?]**   **RUTHIN**
born in Llanderfel, Merioneth
1881 (census: "Age 13; 4 Wern Fechan; Apprentice of Watchmaker")

TOMES   Henry John   **RHYL**
1886 (Dir.: "31a Bedford st")

* TOMKIES   Richard Brown   **[c.1827 - ?]**   **RHYL**
1881 (census) - 1889 (Dir.)   *page 62*

WILCOCKSON Thomas   **DENBIGH**
no date recorded (CWMW)

* WILLIAMS   Hugh   **DENBIGH**
1794 (DPR)   *page 29*

* WILLIAMS   John (I)   **[c.1770 - 1842]**   **DENBIGH**
1802 (DPR) - † 1842 (DPR)   *pages 37, 45-49*

* WILLIAMS   John (II)   **[c.1795 - 1839]**   **DENBIGH**
1822 (Dir.) - † 1839 (DPR)   *pages 37, 45-49*

* WILLIAMS   John Gabriel   **[c.1819 - ?]**   **DENBIGH**
1841 (watch in Welsh Folk Museum) - 1856 (Dir.)
                                       pages 45, 49

* **WILLIAMS**	**John Price**	[c.1859 - ?]		**RUTHIN**
1881 (census)		*page 26*		

WILLIAMS   **Joseph**   [c.1831 - ?]   **DENBIGH**
born in Abergele - married to Sarah
1850 (Dir.: "Henllan") - 1883 (Dir.: "15 Beacon hill")

* **WILLIAMS**   **Owen**   [? - 1756]   **RUTHIN**
† 1756 (RPR)   *page 29*

WILLIAMS   **Thomas**   [c.1811 - ?]   **RUTHIN**
born in Anglesey
1871 (census: "Age 60; 4 Lon Bach, Llanfwrog; Clock Maker")

* **WILSON**   **Gustavus Adolphus**   **DENBIGH**
1815 (DPR)   *pages 42-43*

WOOD   **Alfred**   **RHYL**
1883 (Dir.: "3 High st") - 1886 (Dir.: "16 Bodfor st")

BIBLIOGRAPHY

There are many excellent books on clocks. Some of them are a general introduction to the history of clocks, others specialise in particular makers or particular types of clock, and others again concentrate on specific regions.

Two fine introductions to the history of clocks are:

    1. The History of Clocks and Watches, by Eric Bruton (ISBN 0 7481 0245 0)

    2. The World's Great Clocks and Watches, by Cedric Jagger (ISBN 0 86136 685 9)

In addition, there is a wealth of information in an encyclopedia entitled:

    3. The Country Life International Dictionary of Clocks, edited by Alan Smith (ISBN 0 600 55991 2)

For a detailed discussion on clock styles, in particular the cases, I can recommend:

    4. The Guiness Book of Clocks, by Alan Smith (ISBN 0 85112 225 6)

The long-case clock is well-represented and three titles spring to mind:

    5. The Longcase Clock, by Tom Robinson (ISBN 0 907462 07 3)

    6. Grandfather Clocks and their Cases, by Brian Loomes (ISBN 1 85170 376 4)

    7. English Country Grandfather Clocks, by Richard C.R.Barder (ISBN 1 85170 229 6)

The standard work on the white dial clock is still:

    8. White Dial Clocks, The Complete Guide, by Brian Loomes (ISBN 0 7153 8073 7)

For a good account of a wide variety of 19th and early 20th century clocks, the best book to read is:

    9. The Price Guide to Collectable Clocks, by Alan & Rita Shenton (ISBN 0 907462 66 9)

Finally, for anyone who is keen to research local clockmakers, there is a most useful book on local history which is called exactly that:

    10. Local History, by Philip Riden (ISBN 0 7134 3871 1)

ILLUSTRATIONS

Front cover Brass-faced clock by Richard Griffith of Denbigh (see also figures 17-19)

Figure		
1	Geographical Distribution of Clockmaking in the Vale	2
2	Text of Robert Farr's admission to Freedom of Denbigh	4
3	Clockface styles	11
4	Face of Henry Minshull clock	13
5	Do. (detail of signature)	13
6	Text of sentence passed on John Minshull	15
7	Face of first Minshull clock	16
8	Face of second Minshull clock	16
9	Do. (detail of signature in cartouche)	16
10	Text of William Courter's bill	22
11	Robert G. Joyce - watch-case paper	27
12	Henry Joyce - watch-case paper	27
13	H.E. Joyce - watch-case paper	27
14	Dial of Christopher Sharrock clock	31
15	Do. (detail of signature)	32
16	Do. (detail of seconds ring)	32
17	Dial & hood of R. Griffith brass-faced clock	35
18	Do. (detail of signature)	35
19	Do. (close-up of finial)	35
20	Dial & hood of R. Griffith white-dial clock	36
21	R. Griffith watch	36
22	Portrait of David Griffith (Clwydfardd) aged 92	38
23	Griffith & Son - watch-case paper	41
24	Griffith long-case	41
25	Dial & hood of Jn⁰ Williams white-dial clock	46
26	Dial & hood of J. Williams Jnr white-dial clock	47
27	Do. (close-up of finial)	47
28	Dial & hood of J. Williams white-dial clock	47
29	Do. (close-up of finial)	47
30	J. Williams wall clock	48
31	American wall clock by Keepfer	56
32	Keepfer long-case	56
33	Do. (detail of arch)	56
34	Copy of Keepfer invoice	58
35	Keepfer watch-case paper	60
36	W. Keepfer in front of his shop	60
37	St. Asaph Cathedral clock face	63
38	Trefnant Village post office clock	64
39	Rhyl clock tower	64
40	Old Ruthin Town Hall showing clock	65
41	Ruthin memorial clock	65

Back cover Brass-faced clock by the Minshulls of Denbigh (see also figure 7)